Fiscal and Economic
Implications
of
Strategic Defenses

ABOUT THE BOOK AND AUTHORS

Estimates of the potential costs of alternative strategic defense systems should be an intrinsic element in decisions about whether the United States should develop and deploy these systems, especially because of the "opportunity costs" involved—what the nation would have to forgo in other military or civilian programs or in private resources because of higher taxes or larger deficits.

In this book, Drs. Blechman and Utgoff describe four notional strategic defense systems, each with a different ambitious objective, and then estimate their costs. The first system, Alpha, would seek only to make U.S. nuclear retaliatory forces militarily unattractive to attack. The second, Beta, adds limited protection for the forty-seven most densely populated metropolitan areas of the United States and Canada. Space-based components included in the third and fourth systems, Gamma and Delta, provide comprehensive defenses against Soviet long-range missiles and aircraft; these systems also incorporate options to defend against Soviet intermediate-range missiles. Gamma utilizes interceptor missiles deployed on satelites in low-earth orbits to attack Soviet missiles in their boost phase, while Delta employs chemical lasers on satellites for the same purpose. Assumptions and calculations necessary to make the cost estimates are described in specific terms, and the main determinants of the cost of each system are identified.

In the final section of the book, the authors examine the fiscal consequences of expenditures for strategic defenses in four budgetary contexts: the portion of the defense budget that historically has been allocated for strategic forces, the total defense budget, the federal budget overall, and the nation's economy as a whole. The authors conclude that unless far less expensive ways can be found for building strategic defenses, a decision to deploy such weapons would pose very difficult economic choices for the nation.

Barry M. Blechman is president of Defense Forecasts, Inc., and a fellow of The Johns Hopkins Foreign Policy Institute. He is the author of *U.S. Security in the Twenty-first Century* (Westview, 1986) and coeditor of *International Security Yearbook 1984/85* (Westview, 1985). **Victor A. Utgoff** is a deputy director of the Strategy, Forces and Resources Division of the Institute for Defense Analyses.

NUMBER 12

SAIS
PAPERS IN INTERNATIONAL AFFAIRS

Fiscal and Economic Implications of Strategic Defenses

Barry M. Blechman and Victor A. Utgoff

Routledge
Taylor & Francis Group

NEW YORK AND LONDON

First published 1986 by Westview Press, Inc.

Published 2021 by Routledge
605 Third Avenue, New York, NY 10017
2 Park Square, Milton Park, Abingdon, Oxon OX14 4RN

Routledge is an imprint of the Taylor & Francis Group, an informa business

Copyright © 1986 by The Johns Hopkins Foreign Policy Institute, School of
Advanced International Studies (SAIS)

Library of Congress Catalog Card Number: 86-51129

ISBN 13: 978-0-3670-0656-3 (hbk)
ISBN 13: 978-0-3671-5643-5 (pbk)

CONTENTS_____

ILLUSTRATIONS

TABLES

ACKNOWLEDGMENTS

This book was prepared for a project on "The Military Uses of Space" at the Foreign Policy Institute of The Johns Hopkins University School of Advanced International Studies. The writers are grateful to Dr. Harold Brown, chairman of the institute, and to Dr. Simon Serfaty, its executive director, for their assistance. The project is funded by the Carnegie Corporation of New York.

Suggestions, comments, and reviews of earlier drafts of this book were received from a number of individuals, including: Seymour Deitchman, Joshua Epstein, R.G. Finke, Paul Goree, David Graham, Lou Kaufman, Kevin Lewis, Gerald McNichols, Robin Pirie, and Alan Shaw. We also would like to acknowledge the help of Arnold DiLaura and Margaret Sullivan, who provided research assistance, and Carolyn Finney, for administrative support. Despite all this help, for which we are thankful, the writers remain solely responsible for the final contents of the book and assume full blame for any mistakes, misjudgments, or misconceptions.

EXECUTIVE SUMMARY

ESTIMATES OF THE POTENTIAL COSTS of alternative strategic defense systems should be an intrinsic element in decisions on whether or not the United States should develop and deploy such a system and, if so, which one. Such estimates can identify the likely dominant cost components of any proposed system and thus help decisionmakers in allocating research funds and managerial resources. Such estimates also can provide a sense of the "opportunity costs" that ultimately would be associated with any particular strategic defense system: what the nation would have to forgo—in terms of other federal programs or private resources because of additional taxes or greater deficits—to deploy strategic defenses. With this information in hand, officials and citizens would be in a position to judge whether or not the potential gain from a proposed strategic defense system would be sufficient to merit the associated expenditures and forgone opportunities.

Making such estimates is difficult at this early stage in the research program insofar as the feasibility of alternative technologies is still being explored and no specific design has yet been selected. Questions of technical feasibility as well as probable performance characteristics remain to be resolved. Yet, decisions taken in the near term about the strategic defense initiative (SDI) have consequences for U.S. relations with its allies, for the character of U.S.-Soviet relations and the prospects for arms control, and for the allocation of national scientific and industrial resources. For example, the administration has protected SDI both in allocating the fiscal 1986 budget reductions necessitated by the Gramm-Rudman-Hollings legislation and in its proposals for the fiscal 1987 budget. As such, other defense requirements have been sacrificed. Although relatively small amounts have been involved so far, this process is likely to continue for the next few years and will involve substantially larger figures. Budgetary decisions that favor SDI should not be taken without some empirically derived understanding of the potential costs of strategic defenses, as well as their potential benefits. Economic constraints are as much a part of the reality of national security decisionmaking as are technological limits and geographic barriers. Ignoring these constraints can lead to faulty decisions just as surely as can mistaken assessments of enemy capabilities.

This book is intended to initiate a process of continually improving assessments of the potential costs of strategic defenses. In it, we describe and estimate the development, procurement, and operating costs of four notional strategic defense systems, each of which illustrates one means of achieving a successively more ambitious defense objective. To make the estimates, it was necessary to describe in specific terms the characteristics and size of each element—sensors, weapons, and command subsystems—that would constitute each system. In doing this, we are not suggesting that these specific notional designs necessarily would be the most efficient means of achieving the postulated objective, nor even that all the technical solutions postulated would prove to be practical or effective; such judgments would be premature. The notional systems are simply four designs that conceivably could be developed in pursuit of national objectives.

Our estimates, admittedly, are rough; we have used simple techniques of assessing the numbers of each type of component required to fulfill the objectives of each system and the most straightforward means available to proceed from those numbers to each system's cost. We have sought to describe our assumptions and explain our calculations in sufficient detail so that readers can replicate the exercise and even create variants of the four specific systems outlined in the book. We also have identified the main determinants of the cost of each system and carried out analyses of the sensitivity of the overall systems' costs to variations in those specific assumptions.

COSTS OF NOTIONAL STRATEGIC DEFENSE SYSTEMS

The four systems, their objectives, and their costs are outlined in Table E-1 and summarized below.

Notional System Alpha

The purpose of the Alpha system would be modest: to make U.S. nuclear retaliatory forces militarily unattractive to attack. In operational terms, the system would be designed so that an aggressor nation would have to utilize more nuclear warheads to attack any location of U.S. strategic offensive forces than it could expect to destroy at that location. If this objective could be accomplished, an adversary would have less incentive to initiate a nuclear attack, and the likelihood of a nuclear exchange—even in a crisis—should be reduced.

The system would employ only components that the United States could build in the near future. If a decision was taken in 1987 to develop and deploy the Alpha system, it could be operational in the very early years of the next century.

For defense against ballistic missiles, the system would employ two types of interceptors based on the ground near sites of U.S. offensive forces. Longer-range interceptors, controlled by radars deployed on specially built aircraft, would destroy incoming offensive reentry vehicles (RVs) just as they began to enter the upper atmosphere. Faster, shorter-range interceptors would attack those RVs that managed to penetrate this upper defense layer.

The Alpha system also would include air defenses, so that the Soviets would not perceive options for attacking U.S. offensive forces with manned aircraft or cruise missiles. This component would consist of three elements: (1) early warning aircraft, which would be maintained continuously in the air, close enough to Soviet territory to provide two hours warning of an attack, (2) an outer barrier of surveillance aircraft equipped with long-range air-to-air missiles, and (3) an inner barrier of shorter-range interceptor aircraft guided by AWACS-type aircraft.

Developing, procuring, and operating the Alpha system for ten years would cost roughly $160 billion. (All cost figures in the book are expressed in constant 1987 prices.) Roughly one-fourth of this

TABLE E-1
Characteristics and Costs of the Four Notional Strategic Defense Systems
(all costs in constant fiscal 1987 dollars)

Alpha

Purpose: To make U.S. nuclear retaliatory forces militarily unattractive to attack

Components: Two types of ground-based interceptors deployed at sites of offensive forces for ballistic missile defense; early warning aircraft, plus long-range surveillance aircraft armed with air-to-air missiles, plus shorter-range interceptors for air defense

Full operational capability: 2005; *cost to build and to operate for ten years:* $160 billion; *average annual expenditures during peak ten years:* $10 billion

Beta

Purpose: In addition to that of Alpha, to provide limited protection of the forty-seven most densely populated metropolitan areas of the United States and Canada

Components: Alpha plus additional ground-based, long-range interceptors within areas of relatively high population densities

Full operational capability: 2005; *cost to build and to operate for ten years:* $170 billion; *average annual expenditures during peak ten years:* $11 billion

Gamma

Purpose: To provide comprehensive defense against Soviet long-range missiles and aircraft, plus option to defend against intermediate-range missiles

Components: Beta plus space-based component consisting of interceptor missiles deployed on satellites in low-earth orbits controlled by battle management satellites in 5,000-km. orbits

Full operational capability: 2012; *cost to build and to operate for ten years:* $770 billion; *average annual expenditures during peak ten years:* $44 billion

Delta

Purpose: To provide comprehensive defense against Soviet long-range missiles and aircraft, plus option to defend against intermediate-range missiles

Components: Beta plus chemical lasers in low-earth orbits controlled by battle management satellites in 5,000-km. orbits

Full operational capability: 2020; *cost to build and to operate for ten years:* $670 billion; *average annual expenditures during peak ten years:* $37 billion

amount would be invested in the interceptor missiles for the upper ballistic missile defense layer; one-fifth would be spent on the armed surveillance aircraft in the outer air-defense barrier.

Scaling back the objectives of the Alpha system could result in substantial cost savings. If the United States decided to ignore the threat posed by Soviet bombers and cruise missiles, for example, strategic offensive forces could be made unattractive to attack by ballistic missiles alone for between $40 and $50 billion. If the U.S. objective was even more narrow—to defend only its land-based ballistic missiles from Soviet ballistic missiles—the cost would be about $30 billion.

Notional System Beta

The Beta system would add a light area defense capability to the defenses of the Alpha system. This addition would provide limited protection for the forty-seven most densely populated metropolitan areas of the United States and Canada.

The Beta system would utilize the same types of components as the Alpha system and also could be deployed in the early years of the next century. It would cost $170 billion—only $10 billion more than the smaller system. The degree of protection afforded to the population would be limited, however, as Beta could be easily overwhelmed or evaded.

Notional System Gamma

The Gamma system would be intended to protect the entire population of the United States and Canada from Soviet ballistic missiles and aircraft. To accomplish this, it would add a space-based component to the ground-based and airborne components utilized in the Alpha and Beta systems. The space-based component would consist of interceptor missiles clustered on "battle satellites" in low-earth orbits, which would attack Soviet offensive missiles in their "boost phase" before the Soviets could deploy individual reentry vehicles and decoys. Battle management satellites in 5,000-km.-high orbits would detect the launch of Soviet missiles and track their trajectories, then allocate the interceptor missiles to specific targets and guide them to the vicinity of their targets. As the interceptors neared their targets, on-board guidance systems would cause them to collide with the offensive missile at very high speeds and destroy them. A second wave of interceptors would be launched 10 seconds after the initial wave and, late in their flight paths, these interceptors would be assigned to those boosters that had survived the initial wave of attacks.

The technologies required by the Gamma system are well known, and the system could be developed rapidly. The interceptor, however, would present certain serious challenges. If a decision to deploy Gamma was taken in 1987, it probably could be fully operational around 2012.

If the Gamma system was designed to protect against Soviet ICBMs and SLBMs only, along with air attacks against the United States and Canada, it would cost about $630 billion to develop, procure, and operate for ten years. The system also could be "thickened" to provide protection against Soviet intermediate-range missiles as well; this would raise its cost to a total of $770 billion. If these expenditures were spread over time in a pattern similar to those typical of major weapon acquisition schedules, deploying the Gamma system would require annual expenditures on the order of $44 billion during the peak ten years of funding, 2005 through 2014.

Notional System Delta

The Delta system also would be designed to provide a comprehensive defense, and it would add a space-based component intended to destroy Soviet offensive missiles in the boost phase to accomplish this objective. In place of interceptor rockets, however, the Delta system would make use of chemical lasers deployed on battle satellites in low-earth orbits to attack offensive missiles in the boost phase. These weapons would be directed by battle management satellites comparable to those used in the Gamma system. The laser system probably could not be deployed fully until around the year 2020.

Estimates suggest that it would cost about $550 billion to develop, procure, and operate the Delta system for ten years. If the system was thickened to protect against Soviet intermediate-range missiles, the cost would rise to about $670 billion. Spread over typical weapon acquisition patterns of expenditures, the system would cost about

$37 billion per year during its peak ten years of funding, 2011 through 2020.

In contemplating the costs and benefits of deploying comprehensive strategic defenses, one must analyze a situation in which both the United States and the USSR had taken such a step. Under such circumstances the keystone of NATO strategy—the threat, if conventional defenses fail, to escalate any conflict to a level that would involve the use of long-range nuclear forces—would no longer be credible. Thus, if the United States wished to deploy strategic defenses and yet also to maintain its current international security commitments, this logically would require incremental expenditures to strengthen NATO's conventional and short-range nuclear capabilities as well as to deploy defenses against Soviet short-range missiles and aircraft. These complementary costs, which presumably would be shared by all members of the Alliance, could reach an additional $160 billion.

(In discussing the fiscal and economic consequences of the two comprehensive systems, Gamma and Delta, we have assumed that while the United States would bear the cost of thickening those systems to protect against Soviet intermediate-range missiles, the complementary costs of improving NATO's ground-based defenses, short-range nuclear forces, and conventional forces would be borne by the allies. The additional $160 billion is thus excluded from our system cost estimates.)

CRITICAL ASSUMPTIONS

The most important assumptions made in the course of developing these cost estimates are listed in Table 12.

Optimistically, we have assumed that it would be possible to build interceptor missiles—both the two types of ground-based interceptors utilized in all four systems and the space-based interceptors utilized in the Gamma system—with characteristics that would give them a probability of .9 of destroying their targets in any single engagement. If such an accomplishment was not feasible, the costs of building defense systems whose overall performance approximated that described, would rise substantially. If, for example, the single-shot kill probabilities of all three types of interceptors was .8, the cost of the Gamma system would rise by 30 percent, bringing the cost of the thickened version to just about $1 trillion.

A second critical, optimistic assumption is that relatively inexpensive steps would be sufficient to protect the space-based components of the Gamma and Delta systems from Soviet antisatellite efforts. We have assumed that the United States would harden its satellites to protect them from directed energy systems comparable to those that now exist, and that five decoy satellites would be deployed for every real battle satellite. These two steps would add relatively little to the cost of the systems. If the Soviet Union continues to develop antisatellite capabilities, however, it might be in a position to impose much greater burdens on any U.S. strategic defense system employing space-based assets.

A third critical assumption, which may be either optimistic or pessimistic, is that key components of all four systems—particularly interceptor missiles and battle satellites—would be built in sufficient numbers in precisely the same design so that savings would be possible because of the learning effects of manufacturing experience. We have assumed a 90 percent learning curve, meaning that the marginal costs of affected components would decrease by 10 percent every time the quantity to be procured was doubled. If the assumed learning curve was even only slightly higher or lower, it could make a big difference in costs. If, for example, an 85 percent learning curve was used for the interceptor missiles employed by the thickened Gamma system, its estimated cost would decline by more than 25 percent, or $200 billion.

A fourth critical, and probably pessimistic, assumption concerns the characteristics of Soviet missiles. We have assumed that during the fifteen or more years from the time the United States decided to develop a strategic defense system until such a system became operational, the USSR could replace its force of offensive missiles and tailor their characteristics to make them less vulnerable to the type of defense that the United States had selected. Thus, in assessing the Gamma and Delta systems we have assumed that Soviet land-based missiles would be deployed in an area roughly one-third the size of that in which they are now deployed, that Soviet boost times would be reduced substantially, and that Soviet missiles would be hardened to a degree against directed energy weapons. Although these assumptions are not nearly as pessimistic as could be, each change would impose a penalty on the USSR that it might not choose to accept.

STRATEGIC DEFENSE COSTS IN PERSPECTIVE

The United States clearly could afford to deploy a strategic defense system if it chose to do so. The most expensive notional system, Gamma, would entail incremental annual expenditures on the order of $44 billion during its ten most demanding years. This is obviously a large amount of money; more than a 15 percent real increase in the current level of defense outlays. In other terms, it would represent a commitment of roughly 1 percent of the nation's resources for a sustained period of time. Such a commitment, however, would raise defense expenditures to about only 7 percent of the gross national product (GNP), a figure that has been far exceeded in wartime and matched or exceeded during all but a few of the peacetime years from 1945 to 1970. (Since 1970 U.S. defense spending has varied between about 5 and 6 percent of GNP.)

Still, the pertinent question is not whether the country could afford strategic defense theoretically; it is what the nation would have to give up to do so. Dollars allocated for strategic defenses would not be available for other defense needs or for federal civilian programs. On the other hand, if all other federal programs were protected, the cost of strategic defenses would have to be taken from private resources by raising taxes or deficit financing. Either way, someone must pay the price.

This book explores the opportunity costs of the four notional strategic defense systems by examining their budgetary requirements in the contexts of (1) traditional levels of spending on strategic forces, (2) the defense budget overall, (3) the total federal budget, and (4) the U.S. economy in aggregate. Among other things, it concludes

(1) Strategic defenses could not be financed within historical levels of spending on strategic forces unless: (a) that portion of the defense budget was increased to between $40 and $50 billion (a 60–100 percent increase from current levels), which was the level of expenditures (in current prices) on strategic offensive and

defensive forces in the 1950s; (b) there were no demands for spending on strategic offensive forces beyond the levels envisioned in the administration's current modernization program; and (c) spending on strategic defenses was held to the levels envisioned for the Alpha or Beta systems (roughly $10 billion per year). Neither of the systems employing space-based components could be financed at historical levels of spending for strategic forces under any realistic circumstances.

(2) Factoring out the incremental costs of the wars in Korea and Vietnam, the defense budget has risen, on average, close to 2 percent per year in real terms since World War II. Assuming that this long-term growth rate will continue, given competing demands for additional defense resources, it is unlikely that there will be a sufficient margin to pay for a comprehensive strategic defense system—or even to expand the strategic portion of the budget to pay for the Alpha or Beta systems—unless trade-offs are made with other defense requirements. As examples of such trade-offs, the $160 billion ten-year systems cost of the Alpha system is comparable to the cost of 8 aircraft carrier battle groups, or 27 wings of F–15 fighters, or 14 armored divisions. Annual expenditures for the Gamma system would require even more draconian trade-offs. Those costs are roughly comparable to what either the U.S. Navy or Air Force currently invests in all weapons development and procurement; they would be twice what the Army now spends for such purposes.

(3) Alternatively, defense spending could be raised beyond its historical trend to finance strategic defenses. This could be accomplished without increasing the government's share of national resources by cutting spending on federal civilian programs. Annual peak ten-year expenditures required for the Alpha system would be roughly the same as current annual federal outlays for higher education, for example. The $37 and $44 billion annual peak ten-year expenditures required for the Delta and Gamma systems, respectively, would compare to the $25 billion expended in 1986 on farm income stabilization, the $30 billion spent for health care services, and the $71 billion expended for Medicare. All told, funding the Delta system during its peak ten years would require cutting about one-fifth of the current $180 billion in so-called ''discretionary non-defense federal spending.''

Federal revenues will increase with economic growth, of course, even in the absence of tax increases, but the aging of the U.S. population combined with other proposed uses for federal expenditures, such as infrastructure renewal and improvement of the educational system, will impose strong, competing demands.

(4) Increasing federal revenues to pay for strategic defenses will not receive serious attention this year, but it could be considered in the future. Financing the Gamma system in this way, for example, would require roughly an 11 percent increase in federal revenues from individual income taxes (based on 1985 returns). For the average family earning between $30,000 and $50,000 per year, this would mean an increase of about $570 per year in their tax bill. Alternatively, under the current tax code, the system could be financed by raising revenues from corporate income taxes by about 50 percent, or by roughly doubling the income from excise taxes. Of course, any increase in federal taxes would mean reduced individual consumption or savings and potential effects on economic growth, employment, and inflation.

(5) Raising the government's share of national resources to finance a strategic defense system would not necessarily have adverse economic effects. From 1979 through 1983, for example, total federal outlays as a share of GNP rose by 3.8 percentage points due both to the defense buildup and the cut in tax revenues, yet a combination of monetary policies and favorable international economic conditions avoided the negative effects that had been associated with comparable growth in the federal share during the Vietnam War. Generally speaking, the macroeconomic effects of deploying strategic defenses are likely to be dominated by fiscal and monetary policies and international economic circumstances.

(6) Specific industrial and scientific sectors might be affected more significantly. An undertaking on the scale of the Gamma system could strongly impact on the availability and price of certain kinds of scientists and engineers, computer programmers, and other specialists. It also could distort markets for specific types of raw materials and manufactured goods. Given sufficient lead times, however, federal interventions to encourage growth in relevant occupations, as well as effective cooperation between the federal government and impacted industries, could likely minimize any adverse effects.

INTRODUCTION

A DECISION BY THE UNITED STATES to develop and deploy a system to defend U.S. territory against ballistic missiles would represent not only a major departure in U.S. foreign policy and defense strategy, but also a significant commitment of the nation's resources. Yet, in the extensive and sometimes heated debate since March 1983, when President Reagan announced his intention to explore the possibility that new technologies might make missile defenses feasible, relatively little attention has been paid either to the cost of a strategic defense system or the economic implications of such a decision. Few observers have addressed the issue; fewer have been willing to make specific cost estimates.

Of those who have been willing to address the cost question, some have been optimistic that the expense would be relatively low. General Daniel Graham (USAF-ret.), for example, speaking in 1984 on behalf of the High Frontier (a private organization that has advocated the deployment of missile defenses for many years) estimated that a three-layered, space-based system utilizing kinetic energy weapons would cost $15 to $30 billion.[1] Similarly, in 1985, writing in favor of strategic defenses in *The New York Times Magazine*, Zbigniew Brzezinski, Robert Jastrow, and Max Kampelman estimated that a two-layered defense system could be deployed by the early 1990s for something, "...in the neighborhood of $60 billion."[2]

More frequently, estimates of the cost of strategic defenses have tended to be substantially higher. For example, the Union of

Concerned Scientists (a private organization that has been very critical of the initiative) estimated that the cost of a space-based, directed energy system intended to counter a specific Soviet missile force would be $1 trillion.[3] Similarly, based on what it said were "general historical relationships between pre-full-scale development expenditures and total production costs," the Council on Economic Priorities—also critical of the initiative—estimated the cost of a comprehensive strategic defense system to be "roughly $400 to $800 billion."[4] These latter assessments approximate the most common "back-of-the-envelope estimates" of experienced defense officials and experts; figures between $500 billion and $1 trillion are the ones most frequently encountered.

So far, administration officials have been reluctant to put a price tag on strategic defenses. One exception was the reported estimate by an unnamed official that the space-based component of a multilayered strategic defense system should cost between $87 and $174 billion.[5]

Most often, government representatives have argued that it is premature to estimate the cost of strategic defenses. General James A. Abrahamson, director of the SDI Office, for example, reportedly said that "it would not be responsible" to estimate the cost of strategic defenses because of technological and other uncertainties. Instead, he is reported to have said that the administration intends to announce "cost goals" for the system, "an idealized estimate of what such a defense might cost if prices for high technology could be sharply lowered, not an extrapolation or estimate based on current prices."[6]

Such an idealistic approach may neglect a set of considerations that should be fundamental to the nation's decision about whether or not to develop strategic defenses. First of all, such an approach makes it difficult to identify the likely dominant cost components of any system, an assessment that could be helpful in determining research and administrative priorities as well as in designing the most efficient architecture.

More important, the absence of empirically based cost estimates means that near- and mid-term decisions are not informed by what should be an important criterion. Unfortunately, it is not possible to explore defensive technologies in an abstract context. Decisions taken in the near term about the strategic defense initiative (SDI), among other things, have consequences for U.S. relations with its allies, for the character of U.S.-Soviet relations and the prospects for arms control, and for the allocation of national scientific and industrial resources.

In allocating the reductions in fiscal 1986 defense spending required by the Gramm-Rudman-Hollings legislation, for example, the administration chose to protect the Strategic Defense Initiative, among other programs, meaning that unprotected research and development projects had to accommodate disproportionate reductions. Similarly, the administration instructed the armed services to protect SDI when constructing the fiscal 1987 budget, and it is following suit in its lobbying during this year's congressional review of the 1987 request. While the amounts involved in these decisions so far have been relatively small, they are slated to grow impressively as the exploratory research program progresses; potentially, other defense requirements could be seriously affected within this decade.

Such decisions should not be taken without an at least approximate understanding of the potential costs and benefits of the ultimate objective—the defensive system that eventually might be constructed. And an essential element in any such evaluation should be a continuing assessment of any defensive system's likely opportunity cost: what the nation might have to forgo to deploy it. These estimates can, and should, incorporate assessments of potential reductions in the cost of different types of equipment based on improved technologies and industrial techniques. But they must have some empirical foundation; estimates derived solely from an idealized conception of what various items "should" cost could prove to be misleading. Cost estimates more appropriately should be based on empirically derived analyses of what defensive systems are most likely to cost, and they should be updated continually. For it is only with realistic assessments of potential costs, as well as of potential technological capabilities and political and military consequences, that the nation will be able to decide rationally about the optimum course of action in both the near, and the more distant, future.

It is the potential magnitude of the expense of strategic defense systems, of course, that makes realistic estimates of their costs so essential. If the many cursory estimates that comprehensive missile defenses might cost as much as $500 billion or $1 trillion have any plausibility, the question of systems' cost and their potential economic consequences might be no less important than such fundamental issues as technological feasibility and international impact. Rarely has the nation deliberately allocated such large sums for any single military purpose. Programs that we consider to have been major efforts pale by comparison. NASA's entire budget during the fiscal years 1960–69, when the nation made a concerted and ultimately successful effort to land men on the moon, for example, totaled only about $120 billion when expressed in 1987 prices. Even wars have cost less than the potential cost of comprehensive strategic defenses; at 1987 prices, for example, the incremental cost of the war in Vietnam from fiscal 1965 through 1972 came to about $300 billion. Total U.S. defense expenditures during the three years of the Korean War came to about $650 billion. (We look extensively at comparisons between the potential costs of strategic defenses and other major national programs in chapter 9.)[7]

The key question is how much the United States might have to allocate for strategic defenses each year and, consequently, what it would have to give up to make such a sum available. Developing and constructing a comprehensive strategic defense system would no doubt take a sustained period of time—fifteen to twenty-five years is a common estimate—and the system would then have to be maintained and operated and then improved continually. As the United States has no such system now, and has expended only very small amounts on strategic defenses for many years, for the most part these sums would represent new demands on the nation's resources each year.

If costs did total $500 billion to $1 trillion, they would translate into average annual expenditures of $25 to $50 billion over a twenty-year period. In very rough terms, this would represent something like a 10 to 20 percent real increase in the current level of defense outlays. It would double or triple what the country now spends for strategic forces, both offensive and defensive, which is already relatively high by the standards of the past twenty years.

Expenditures of this magnitude should certainly be possible for a nation as wealthy as the United States. Adding even $50 billion to planned fiscal 1987 defense outlays would raise defense expenditures to about only 7 percent of the projected gross national product (GNP)—a figure that has been far exceeded in wartime and matched or exceeded for all but a few peacetime years from 1945 to 1970. For the past fifteen years, however, defense spending has fluctuated more or less between 5 and 6 percent of the nation's resources. A decision to increase that share by a full percentage point for a sustained period of time for the one specific purpose of building strategic defenses would represent a significant departure in national priorities.

Before determining to set off on such a course, the United States should look hard at the potential consequences of such expenditures. Economic constraints are as much a part of the reality of national security decisionmaking as are technological limits and geographic barriers. Ignoring economic considerations can lead to faulty decisions and dangers to the nation's security just as surely as can mistaken assessments of enemy capabilities.

Conceivably, some will judge the opportunity costs of building missile defenses to be too high relative to the potential gain; they may consider such incremental expenditures likely to affect adversely the nation's ability to meet other defense needs or to satisfy requirements for domestic programs, or they may believe that such new federal expenditures would adversely affect the country's overall economic performance. Conversely, others may see a program of this magnitude as likely to serve primarily as an economic stimulus, helping to create jobs and spinning off technologies that could improve productivity and the nation's global competitive position in a host of related industries. Still others may come to believe that such expenditures on missile defenses could be accommodated without major economic dislocations one way or the other, perhaps as a result of reductions that would be made possible in other defense programs or because of possible changes in fiscal or monetary programs. Regardless of which assessment ultimately gains the widest support, it seems clear that evaluations of the costs of building missile defenses and their prospective economic consequences are important topics for national debate.

This book is intended to initiate such a discussion, to begin a process of continually improving assessments of the potential costs of strategic defenses. Admittedly, the estimates it provides are rough, and they are based on technical approaches that are likely to differ in many ways from those that ultimately would be adopted should a system of strategic defenses be built someday. Therefore, the actual costs of the strategic defenses selected and constructed by the nation may differ significantly from the estimates made in this book.

This potential discrepancy is no different a problem in principle from that normally faced when estimating the cost of any large project of a type that has never before been accomplished. The first cost estimates for any new type of aircraft, ship, or space system are usually based on plans that will be revised substantially; as a project proceeds it is not unusual for major changes to be made in technical approaches to solve the problems that are revealed in developing and constructing the system. The redesign of the Apollo spacecraft after its disastrous fire is an example of such a major change in technical approach taken well after a commitment had been made to proceed with a very large project.

One could argue, of course, that large projects should not be started until designs have been finalized and the costs established with great precision. The wonders of computer-aided design systems and elaborate simulations notwithstanding, this approach is impractical. The best of plans for a complex undertaking employing new technologies cannot provide a basis for understanding a new system that is nearly as good as the actual experience of trying to build it.

Large-scale high-technology projects inevitably must be started on the basis of engineering plans and cost estimates incorporating significant uncertainties. This is all the more true when the most advanced emerging technologies must be employed to have any reasonable prospect of achieving an effective system, as is often required when an opponent can be expected to employ its best technologies to counter the effectiveness of the adversary's system.

Thus, though the strategic defense systems considered below will surely differ from those that might ultimately be built, they are reasonable, first detailed estimates for a set of plausible alternatives. As the investigation of alternative designs for strategic defenses proceeds, additional and more refined estimates of costs should be made.

In estimating strategic defense costs, we have had to make many assumptions regarding such questions as the kinds of technologies to be employed in solving problems, how well the chosen approaches might work, counteractions that the Soviet Union might take in designing and deploying the offensive forces that would face U.S. defenses, and so forth. These assumptions are highlighted in the body of the book; the effects of the most important assumptions are summarized and discussed toward its end.

In general, we regard the assumptions we have made to be optimistic for the United States, and this judgment is concurred in by nearly all of the many individuals who reviewed previous drafts of this book. To the extent that our technical assumptions are optimistic, the cost estimates should be considered optimistic as well. On the other hand, there is always the possibility of a genuine technical breakthrough that could lead to significantly less costly strategic defense systems than those described here. We discuss some of these possibilities in chapter 6.

In developing the cost estimates, we include comprehensive and detailed descriptions of their derivation—both the assumptions and the calculations upon which they have been constructed—so that readers can determine precisely the degree to which they agree or disagree with our assessments and identify the sources of any difference of opinion. We would welcome comments on the book, particularly its methodologies and assumptions, as well as independent appraisals of the prospective cost and economic consequences of strategic defenses.

The book is divided into two parts. In the first, we estimate the costs of four notional strategic defense systems intended to achieve alternative national objectives. In the second, we set these estimates in perspective, describing the potential trade-offs between expenditures of this magnitude and alternative uses, private and public, of the nation's resources.

PART 1

COSTS OF NOTIONAL STRATEGIC DEFENSE SYSTEMS

IN THIS SECTION WE ESTIMATE the costs of developing, building, and operating four notional systems for strategic defense. The four systems illustrate means of achieving successively more ambitious defense objectives.

To make the estimates, it was necessary to describe each notional system in relatively specific terms, determining the rough characteristics and size of the components—sensors, weapons, and command subsystems—that would constitute each alternative. These assessments are not intended to suggest that the specific notional designs would necessarily be the most efficient means of achieving the postulated objectives, nor even that all the technical solutions suggested would prove to be practical and effective. It is too early to reach such judgments in many cases, and the number of detailed evaluations that would have to go into the specific design of any strategic defense system is too large for our meager resources to deal with definitively. The estimates are intended, and should be taken, solely as illustrations.

In making the estimates, we employed very simple techniques for assessing the numbers of each type of component required to fulfill the objectives of each system and the most straightforward means available to proceed from these numbers to each component's cost. More sophisticated methods are available for building up cost estimates, but they are more difficult to make use of and to follow and in any event are not justified in view of the fundamental uncertainties that exist concerning parameters of individual weapons and sensors. The details of alternative system architectures—and thus more finely honed cost estimates—will become important only as alternative means of achieving a common set of objectives reach a sufficient degree of technological maturity to permit more reliable evaluations.

We have taken pains both to identify the more important unsolved technical problems in the specific system architectures that we use to illustrate potential costs and to explain the reasons for the design choices incorporated in each of our alternatives. These explanations should permit readers to make their own assessments of the feasibility of each notional system. We also have sought to describe our assumptions and explain our calculations in sufficient detail so that readers can replicate the exercise and, indeed, create variants of the four specific systems we describe. Such manipulations of our calculations can provide insights into what are the main determinants of the overall costs of alternative types of strategic defense systems. Such analyses also would permit very rough calculations of the sensitivity of overall system costs to variations in such key parameters as the size of the attack the system is supposed to defeat.

In addition to estimating the cost of each notional strategic defense system, we have estimated the costs of complementary actions that logically would be implemented as part of an overall U.S. military strategy intended to achieve the set of strategic objectives associated with each alternative. The costs of complementary forces are reported separately so that readers can adjust them as they see fit or even ignore them in evaluating the economic implications of deploying missile defenses; reasonable arguments can be made for both including and excluding the costs of complementary systems in such assessments.

By illustrating the potential costs of notional strategic defense systems in these approximate terms, we hope to make possible consideration of what should be an important element in the national decision on whether to pursue strategic missile defenses and, if so, for which purposes.

1.
NOTIONAL STRATEGIC
DEFENSE SYSTEM ALPHA
Defense of Nuclear Retaliatory Forces

T HE FIRST SYSTEM WOULD EMPLOY only components that the United States could build today or in the near future. If a decision was taken in 1987 to build the Alpha system, the system could be completely designed and developed by about 1995 and fully deployed in the early years of the next century. The system would have one design objective: to make U.S. strategic nuclear retaliatory forces militarily unattractive to attack. Theoretically, achievement of this objective could be an effective means of stabilizing the nuclear balance and reducing the risk of nuclear war. It would constitute, however, only a modest step toward President Reagan's goal of "rendering nuclear weapons impotent and obsolete."

In operational terms, the design objective of the Alpha system would be to raise the cost of attacking the bases at which U.S. strategic nuclear forces are located to the point at which more nuclear weapons would be required to destroy each base than could be expected to be found at the base at the time the attack was executed. If this objective could be accomplished, the relative potential value of a preemptive strike against U.S. nuclear retaliatory forces would be reduced and, as a result, the stability of the nuclear balance improved and the risk of a nuclear exchange reduced. Potential adversaries would have less incentive to initiate an attack, as they would face the prospect of using up more of their own nuclear weapons in carrying out the attack than they could reasonably expect to destroy.[8]

The Alpha notional strategic defense system would employ combinations of ground-based missile interceptors and radars, airborne surveillance, target acquisition, and battle management systems to defend against ballistic missile attacks. The missiles would include longer-range (high endo-atmospheric or "HEDI") interceptors, organized into an upper layer of defense that would seek to destroy incoming reentry vehicles (RVs) carrying nuclear warheads above an altitude of 100,000 ft., plus faster, shorter-range (low endo-atmospheric or "LEDI") interceptors that would constitute a lower layer of defense to destroy those few RVs that managed to penetrate the upper defense layer. Intercepts in the lower defense layer would be controlled by individual ground-based radars at each location being defended. The actions of the high endo-atmospheric interceptors would be controlled by airborne laser radar systems, assisted by the ground-based radars. The airborne laser radar systems similarly would assist intercept attempts by the LEDI interceptors.

Early warning of the launch of attacking missiles along with preliminary assessments of the size and potential targets of the attack would be provided by existing or planned warning and surveillance satellites in geosynchronous orbits and ground-based, over-the-horizon radars. No incremental expenditures would be required for these systems.

We have chosen not to employ any other space-based components in this first defense system. It seems likely that any space-based system would require a long development period and also would have potential vulnerabilities for which no confident solution appears to have been identified as yet. (Early warning satellites suffer less from this latter disability, as these high-altitude satellites are not within range of any antisatellite system that has been tested so far and, even if they were, an attack on a warning satellite would itself constitute tactical warning.) Means of reducing the potential vulnerabilities of satellite systems are discussed in the context of the Gamma and Delta systems.

The Alpha missile defense system would be complemented by an airborne system to defend against bombers and cruise missiles. High-flying aircraft carrying long-range surveillance systems would continuously patrol a barrier well to the north, and off the coasts, of the United States. These aircraft would provide the advance warning needed to organize an outer barrier of surveillance aircraft equipped with relatively long-range air-to-air missiles. Bombers able to penetrate this outer barrier would be met subsequently by shorter-range interceptor aircraft directed by AWACS-type aircraft.

Airborne, rather than ground-based, air defense systems have been used whenever possible in designing the Alpha system to reduce its cost. Airborne air defense systems have the potential to be massed along those routes actually selected by an enemy for its strongest attacks. Immobile or slow-moving, ground-based systems, on the other hand, would have to be deployed in sufficient numbers along all potential attack corridors. In the defense of a very large area encompassing many well-separated targets, the use of airborne systems could so reduce the number of units required to defend against a threat of a specific size as to more than make up for the greater cost of airborne systems.[9]

It should be noted that although the technologies that we propose to incorporate in the Alpha system are largely in hand, we are postulating the development of certain weapon systems, such as armed surveillance aircraft for air defense, that have been proposed before and turned down by defense officials. Of the reasons why these types of systems have been rejected in the past, perhaps the most important is that such aircraft would be very expensive on a unit basis. They also lack the flexibility to perform any other mission, in contrast to most current fighter aircraft of the United States.

SIZE AND CHARACTERISTICS OF THE ALPHA SYSTEM

The primary targets to be protected by the Alpha strategic defense system would be the bases of U.S. strategic forces. For the purposes of this analysis, every known location of U.S. strategic forces has been considered a "base," including each individual missile silo. In order to assess the level of defense required at each type of strategic forces base, we must first estimate the number of strategic weapons likely to be located at each of the bases at the time an attack could arrive. These data are contained in Table 1.

For each type of facility, the number of weapons expected to be present at the time of an attack is the product of the total number of missile launchers or bombers located at the base and the number

TABLE 1

Average Number of Weapons Located at U.S. Strategic Bases
at the Time an Enemy Attack Could Arrive
(by type of strategic force)

Type of Strategic Force	Total Number of Launchers in Force[a]	Average Weapons per Launcher	Number of Bases[c]	Average Number of Weapons Left at Base
Trident Strategic Submarine	480	8	2	576
B–1 and ATB Bombers	240	12[b]	30	58
MX	50	10	50	10
Minuteman III	550	3	550	3
Midgetman	500	1	3	17

NOTES: Forty percent of the bomber force are presumed to be sufficiently alert to escape from their bases before an attack arrives; 70 percent of the Trident submarines are presumed to be at sea and the remainder are presumed vulnerable to attack; 90 percent of the Midgetman force are presumed to have dispersed and thus to be effectively untargetable. All of the land-based missile force is assumed to ride out any attack.[10]

[a]Numbers of strategic launchers are the writers' projections of the force structure in the year 2005, when the Alpha system would be fully operational. It seems likely that the Poseidon submarine force would be retired by then, for example, as would the Minuteman II force, the latter in favor of Midgetman.

[b]The average number of weapons per bomber is based on an assumption that the B–1 typically carries 12 weapons, its internal capacity, even though additional weapons could be stored externally. Although the ATB is reported to have a smaller payload, its effective operational load could not be too much smaller if the new bomber is to be practical.

[c]A strategic forces "base" is defined as any location of such forces that would provide an individual aim point for an attack. Thus, although all 50 MX might be located at a single air force "base," in the normal usage of the word, they constitute 50 bases in our usage because each silo must be targeted individually.

of weapons deployed on each launcher or bomber, multiplied by the percentage of those types of launchers or bombers that would be maintained at too low an alert rate to escape from the base in the period between receipt of warning of an attack and its arrival (or that would seek to survive by riding out the attack rather than by trying to escape).

Given our assumed design criterion, the level of ballistic missile defenses that should be deployed at each type of strategic forces base is the amount necessary to make it more costly (in terms of expended warheads) to destroy a base than the base's expected value (also expressed in terms of warheads). The most important factors required to establish the size of each base defense are estimates of: (1) the quality of the defense's interceptors, and (2) the average number of false targets or decoys that might have to be attacked to destroy each enemy reentry vehicle. We have assumed that a missile defense system incorporating only one or two layers of defense would not be built unless the interceptors utilized in each layer had relatively high individual probabilities of kill against attacking reentry vehicles. Thus, in designing the Alpha system, we assumed that each interceptor would have a 90 percent chance of destroying its designated target. Planning on the basis of a 90 percent kill probability (a figure that incorporates degradations for limitations on a weapon's reliability as well as theoretical limitations on its effectiveness when it performs as designed) is conceivable, although optimistic; few things as complex as guided missiles perform so well. The sensitivity of the cost estimates to this assumption, among others, is tested in chapter 6.

Regarding false targets, it is clear that the creation of an effective terminal defense can be made more difficult by surrounding each reentry vehicle with lightweight decoys. Unless the real target could be identified at an early enough point, the decoys could force the defense to waste interceptors. The most straightforward means to identify lightweight decoys would be to wait until they began to reenter the atmosphere, at which point the false targets would begin to be stripped away from the heavier reentry vehicles. Heavy decoys could be used effectively to mask reentry vehicles further into the atmosphere, but their greater weight means that they would use up larger amounts of the "throw-weight" of the offensive missiles that otherwise could be used for real weapons.

We have assumed that the Soviet Union would choose not to pay the "throw-weight" penalty required for the use of heavy decoys and the defense would use this "atmospheric filtering" to achieve perfect discrimination between lightweight decoys and real reentry vehicles. (Readers should note that the use of lightweight decoys benefits the attacker even if atmospheric filtering is perfectly effective. By forcing the defense to make its intercept attempts only after the attacker's weapons began to reenter the atmosphere, the defense would be required to employ interceptors with very high acceleration and accordingly greater cost and consequently would have to limit itself to one or two separate ["shoot-look-shoot"] intercept attempts against each target.)

Assuming that a sufficient number of high endo-atmospheric interceptors were positioned at any particular base, the assumed 90 percent kill probability means that for every 10 attackers aimed at a defended target, only 1 would be expected to penetrate the upper defense layer. The layer of low endo-atmospheric interceptors is thus provided to destroy such "leakers."

Based on these sizing criteria, we can now establish the specific configuration of the Alpha system's ballistic missile and air defense components.

Ballistic Missile Defenses

This component of the Alpha system would consist of HEDI and LEDI interceptors, ground-based radars, and aircraft used for the management of intercepts by the upper layer of defense.

Interceptors and ground-based radars. The sizing of a missile defense strong enough to extract a specified "attack price" depends not just upon the effectiveness of the individual interceptors employed, but also upon the doctrine chosen for allocating defending interceptors to attacking warheads and upon the attacker's strategy as well.

If, for example, the attacker could assume that the target would remain in its base long enough for the attacker to employ intelligence systems to support a shoot-look-shoot strategy, the attacker could repeat the attack until it observed that the target had been destroyed. If, on the other hand, the target could vacate its base in much less time than the attacker would probably need to destroy it using a drawn out shoot-look-shoot strategy, the attacker would have greater assurance of success if it fired a single salvo of warheads at the target, with enough warheads in the salvo to achieve an acceptable probability of destroying the target. Combinations of attack strategies also could be used: if, for example, survival of the target could be tolerated long enough to allow intelligence systems the required intervening look, two salvos in a shoot-look-shoot mode could be used.

Relatively sophisticated strategies are also available to a defender in this type of attack. If, for example, the defense hoped to extract a high price for the target being defended and faced an attacker employing a shoot-look-shoot strategy with single warheads, the defender should probably meet the first attackers with salvos of interceptors. This strategy would give a higher probability of stopping the first few attackers, thus reducing the probability of early destruction of the target, a contingency that would waste those interceptors assigned to the target that never got into action. This consideration, however, must be balanced against the fact that firing interceptors in salvos would empty the defender's quiver more quickly. An optimal strategy, thus, would taper down the number of interceptors used in each successive salvo.

A defender confronting a salvo attack would face a different problem. The primary concern would be to make sure that every attacker was met with its fair share of whatever number of interceptors the defense decided to fire at that attacking salvo. The defender also would have to concern itself with the question of how many attacking missiles it might have to face in the course of an extended defense.

In constructing the Alpha defense, we have reflected these considerations by sizing the defense of each target to allow a defense doctrine that would have no more than a 30 percent chance of failing through leakage before it had exhausted its supply of interceptors. We have further assumed that the defense would have no difficulty in executing its firing doctrine and achieving the anticipated probabilities of kill, even if the complete attack arrived in a single salvo. Finally, since the factors that might make a single interceptor miss its target would occasionally cause accompanying interceptors to miss as well, we have subtracted an additional 10 percent from the probability of kill of each interceptor added to a defending salvo. Thus, the second interceptor in a salvo is assigned a probability of kill of .8, the third .7, and so on.

The firing doctrines adopted for the defense given in Table 2 take these considerations into account. The defenses of each type of strategic base have been constructed by providing sufficient interceptors to allow destruction of a number of attackers equal to the number of U.S. warheads remaining at the base, assuming employment of the firing doctrine given in the table. Note that salvo fire is employed only with low endo-atmospheric interceptors. While these interceptors are more expensive than their high endo-atmospheric counterparts, far fewer additional missiles would be needed because they would have to defend against only "leakers" from the upper layer of defenses. The numbers of high endo-atmospheric interceptors, low endo-atmospheric interceptors, and ground-based radars that would be required to defend each type of base are provided in Table 3.

TABLE 2
Firing Doctrines Employed by the Alpha Defense

Number of Warheads at Defended Base	Doctrine for Defense of the Base
1-3	One HEDI per defended warhead
4-35	One HEDI per defended warhead, plus one LEDI for each leaker expected through the upper defense
36-178	One HEDI per defended warhead, plus one 2-missile LEDI salvo for each of one-half of the leakers expected through the upper defense, plus one LEDI for each of the remaining leakers
179-576	One HEDI per defended warhead, plus one 3-missile LEDI salvo for each of one-third the number of leakers expected through the upper defense, plus one 2-missile LEDI salvo for a second third of the expected leakers, plus one LEDI for the remaining third of the leakers

TABLE 3
Ballistic Missile Defense Configurations Required to Make Attacks on U.S. Strategic Forces Bases Unprofitable or Minimally Profitable

Type of Strategic Force	Number of HEDI per Base[a]	Number of LEDI per Base[b]	Number of Radars per Base
Trident Strategic Submarines	576	115	1
B-1 and ATB Bombers	60	10	1
MX	10	2	0.1[c]
Minuteman III	3	0	0
Midgetman	17	3	1

[a]In several cases the numbers of HEDI have been rounded upward by 1 or 2 per base.
[b]The numbers of LEDI have been rounded up by slightly more than 1 or 2 to allow for statistical variations in the expected leakage through the upper layer.
[c]MX missiles are assumed to be located in silos that are close enough together to allow the lower layer of defense for 10 silos to be managed by a single ground-based radar.

Aircraft for battle management. To estimate the number of aircraft carrying laser radars and battle management systems that would be required in the Alpha system to control intercepts by the upper layer, we have made the following assumptions:

(1) The aircraft would operate at an altitude of 35,000 ft.; all intercept attempts by the upper layer would be completed well above this altitude.

(2) Coverage of the entire area of the United States and a 200-nautical-mile-(n.mi.)-wide band of the territory of Canada adjacent to the United States would be required. U.S. strategic forces bases and the additional air bases that would be required for the complementary air defense system (see below) would be distributed

over this entire area. Inclusion of southern Canada seems reasonable, given the precedents set by Canada's longstanding participation in the North American Air Defense Command and other defense arrangements.

(3) In order to guard against the inadvertent loss of aircraft (some nuclear detonations would undoubtedly take place in unexpected places in the course of the battle) and the corresponding collapse of missile defenses in its sector, a backup aircraft would be kept airborne in each sector of the upper defense layer. If both aircraft guarding any particular sector were operating at the time intercepts were required, they would share the work load.

(4) The laser radar/battle management aircraft would be designed to be capable of escaping from their bases upon first warning of an attack sufficiently quickly that it would not be necessary to maintain them on airborne alert.

(5) Taking training and maintenance requirements into account, 40 percent of the laser radar/battle management aircraft would be kept on strip alert, ready to depart their bases at first warning of a possible attack.

Given these assumptions, the first step in estimating the number of radar/battle management aircraft required is to calculate the size of the sector that could be guarded by a single aircraft. We have assumed that a field of view down to 35,000 ft. would allow the airborne laser radar to provide a sufficient assist to the ground-based radars controlling the lower layer of defenses. This field of view approximately equals the area of a circle with a radius equal to the maximum distance at which a point 35,000 ft. above the earth would appear on the horizon to an aircraft at the same altitude, or about 600,000 sq.n.mi. This would be the maximum size of the sector that could be guarded by each aircraft (plus its backup).

It should be possible to station the aircraft in such a way that no more than 30 percent of their maximum coverage area would be wasted in dividing up the necessary total coverage area of approximately 3.3 million sq.n.mi. If so, the laser radar/battle management aircraft would have to cover eight sectors; 16 such aircraft (primaries and backups) would have to be on strip alert at all times. All told, this would require an inventory of 40 laser radar/battle management aircraft for the upper missile defense layer.

One additional calculation must be made to complete the sizing of the ballistic missile defense component of the Alpha strategic defense system. We must estimate the size of the air defense component, as the home bases of the air defense interceptors also would have to be protected from ballistic missile attacks. We thus turn to the air defense component before completing the structure of missile defenses.

Aircraft and Cruise Missile Defenses

In sizing the air defense system, we have assumed that the Soviet Union deploys a force of 200 Bear F and Blackjack strategic bombers carrying an average of 10 cruise missiles each. This represents roughly the same number of long-range bombers that the USSR has maintained in its force posture for many years, but equipage with modern cruise missiles would constitute a major improvement in Soviet bomber capabilities. Because we have assumed the Soviet Union to be the aggressor, we estimate that two-thirds of the bombers would participate in the initial attack.[11]

All told, the current Soviet submarine force could carry about 700 cruise missiles of all types, assuming that attack submarines were also loaded with 1 SS–NX–21 cruise missile per torpedo tube. Only a relatively small portion of this force could be deployed within range of the United States at any one time. The Soviet submarine force is being modernized continually, however, and its cruise missile capabilities are changing rapidly as Oscar-class submarines enter the fleet and development of the SS–NX–21 is completed. Given these changes, it is possible that around the turn of the century a surge effort might allow the Soviet Union to deploy 350 cruise missiles on submarines off U.S. coasts.

We have assumed further that the bombers and cruise missiles constituting the Soviet threat travel at high subsonic speeds. Supersonic aircraft or cruise missiles can be built, of course, but they would be substantially larger and more expensive. (Adjustment of the calculations given below for supersonic attack aircraft or cruise missiles would be straightforward if readers wished to examine the consequences of that possibility.) We have also assumed that the attacking cruise missiles would have sufficient range to make impractical any attempt to intercept the aircraft carrying them before the missiles had been launched.

To construct an effective air defense, the United States would have to deploy a combination of early warning aircraft and armed surveillance aircraft, and it would have to use the 300 shorter-range interceptors, command and control systems, and 10 AWACS aircraft already planned for continental air defenses in the 1990s. In addition, it would have to build additional air bases to house the new air defense aircraft. Each of these elements is discussed below.

Early warning aircraft. The early warning aircraft used by the United States for air defense in the Alpha system would be maintained continuously in the air, sufficiently close to Soviet territory to allow two hours warning of an attack. During this time, the armed surveillance aircraft would be moved into optimal positions to meet the Soviet bombers.

A modified version of the TR-1 reconnaissance aircraft equipped with a long-range, conformal array radar would be suitable for the early warning task. The radar would be used to detect threatening aircraft, provide a rough picture of how large the potential attack might be, and give some indication of where the attacking aircraft appeared to be headed. If a more detailed picture of the attack was needed, the raw radar data would be relayed back to airborne and ground-based command posts in the United States for further processing.

With an operating altitude of 90,000 ft., TR-1s would have an effective range of about 320 n.mi. when used in an early warning role. They would maintain a continuous barrier approximately 1,000 n.mi. to the north of the U.S.-Canada border and stretching well offshore down the U.S. coasts. The total length of this early warning barrier would be roughly 7,500 n.mi.

If we allow a modest overlap in coverage areas, 15 aircraft on station could cover this perimeter. This would require an inventory of approximately 75 aircraft if 20 percent of the inventory could be kept on airborne alert. This is a relatively high alert rate to sustain for a protracted period of time, as continuous airborne operations stress both the hardware and the crews. It has been achieved in the past, however.

Armed surveillance aircraft and air-to-air missiles. In our model the aircraft to be employed for the armed surveillance role would be similar to existing KC-10 cargo and tanker aircraft, but they would be equipped with a powerful air surveillance radar and the equipment needed to allow a team of eight controllers simultaneously to run four intercepts each. Such capabilities should not be difficult to develop. Existing AWACS and E-2C aircraft have comparable capabilities, if the relative sizes of the aircraft are taken into account, and both existing types of aircraft incorporate established technologies.

Each aircraft would be equipped with 100 interceptor missiles with characteristics roughly comparable to existing Phoenix (54C) missiles, but with 50 percent greater range. In the scenario we envision, attacking aircraft would attempt to penetrate the air defense barrier at low altitudes, which would create a cluttered background for the surveillance radar and constrain the range at which the attacking aircraft could be tracked and intercepted to a maximum distance of 150 n.mi. We have assumed that the interceptor missiles would be designed to make intercepts at this maximum range, but, given the cluttered radar background, they would be unlikely to have single-shot probabilities of kill in excess of 70 percent.

When deployed, the barrier of armed surveillance aircraft defending against cruise missiles launched from bombers would be located in a 150–300-n.mi.-deep area just north of the 200-n.mi. strip of southern Canada included in the protected zone. These aircraft would be based in the northern United States and southern Canada and would be maintained on strip alert. In view of training and maintenance requirements, 40 percent of the aircraft dedicated to this role probably could be kept on strip alert.

The aircraft defending against cruise missiles launched from submarines would operate from bases 100–200 n.mi. inland from the U.S. coasts. These aircraft also would be maintained on strip alert unless significant numbers of Soviet submarines moved close enough to the coasts to jeopardize the timely escape of the surveillance aircraft from their bases. At such times the armed surveillance aircraft would be maintained on airborne alert. Because of this requirement, we have assumed that only 20 percent of the armed surveillance aircraft used to defend against submarine-launched cruise missiles would be available at any one time. (We have assumed further that the United States would be capable of detecting and localizing close-in Soviet submarines to the degree required to provide sufficient warning without developing and acquiring antisubmarine sensors other than those already existing or planned.)

The air defense barrier, which would operate far forward of the targets to be defended, would face the entire Soviet bomber-launched cruise missile attack, rather than the portion of the attack that might be directed at a specific target. The number of weapons the air defense should be required to destroy in the overall Soviet air attack should equal either the total number of U.S. strategic warheads that would remain at their bases when the attack arrived or the number of cruise missiles in the entire attacking force, whichever was greater.

Based on Table 1, the total number of U.S. weapons that might remain at their bases would be either 5,093 or 2,943, depending on whether silo-based missiles were counted. It might be unrealistic to include intercontinental ballistic missiles (ICBMs) as potential targets of an air attack, as they almost surely would be launched from their silos before the bomber-launched cruise missiles could arrive or, at least, as the attacker would have to assume that they would escape. This issue is unimportant, however, since in either case the total number of weapons remaining at the U.S. bases would exceed the roughly 1,700 Soviet bomber- and submarine-launched cruise missiles that we have assumed would constitute the attack. The criterion for sizing the air defense system therefore must be simply the ability to stop as many of the attacking aircraft as possible.

We have assumed that the range capabilities of the bomber-launched cruise missiles would not be great enough to allow lateral moves in excess of 1,000 n.mi. after penetration of the armed surveillance aircraft barrier. Given this assumption, the two-hour warning time provided by the early warning aircraft previously mentioned, the division of Trident submarine bases between the east and west coasts, and the wide distribution of U.S. bomber bases would guarantee an east-west distribution of the attacking bomber-launched cruise missiles that should allow the United States to mass its armed surveillance aircraft exactly where they were needed. In other words, virtually all armed surveillance aircraft on strip alert should be able to participate in the air defense battle.

We further have assumed that the force of interceptor and AWACS aircraft, which is already planned for U.S. air defenses, also would be available in the event of a decision to build the Alpha system and that it would participate in the air defense battle. This force would be used to back up the armed surveillance aircraft barrier. We have assumed that 70 percent of this force would be sufficiently alert to make a contribution and that each of the interceptor aircraft would participate in a sufficient number of intercept attempts to allow it to employ, on average, one load of air-to-air missiles, each fired independently at an attacker.

Given these assumptions, the already planned force of interceptors would be expected to destroy a number of attackers equal to the product of the number of defending aircraft (300), the assumed alert rate (.7), the number of air-to-air missiles carried by each aircraft (4), and the assumed single-shot probability of kill of the missile (.7). The defender could expect a kill of approximately 600 of the attacking cruise missiles, which we have assumed would be divided proportionately between the bomber-launched and submarine-launched weapons.

Thus, 1,100 bomber- and submarine-launched cruise missiles would have to be destroyed by the armed surveillance aircraft in the outer barrier. The number of armed surveillance aircraft required can then be estimated by dividing this total by the expected number of kills per defending aircraft. In calculating the latter, we note that with (1) each surveillance aircraft capable of thirty-two simultaneous intercepts, (2) subsonic attackers, (3) intercepts possible out to a range of 150 n.mi., and (4) a defending air-to-air missile capable of flying at more than five times the speed of the attacking cruise missiles, the defending surveillance aircraft should be able to fire all their missiles effectively. The expected kills per surveillance aircraft would then be the product of the number of air-to-air missiles carried by each defender (100) and the single-shot probability of kill of the air-to-air missile (.7), which is 70 expected kills per defending aircraft. A total of 16 alert surveillance aircraft would thus be required to destroy the 1,100 cruise missiles that would not be destroyed by already planned air defense forces.

Before settling on this total requirement, however, we must ensure that the force also would be large enough to guard the entire U.S. defense perimeter; otherwise, there would be attack strategies available to Soviet planners that could defeat the U.S. defense. Of the 16 required alert surveillance aircraft, approximately 13 would be needed to defeat the bomber-launched cruise missile attack. If some overlap was allowed between the coverage areas of adjacent aircraft, these 13 aircraft could form a continuous barrier roughly 3,100 n.mi. in length—a figure less than the 4,500-n.mi.-long perimeter that must be defended. To allow coverage of the entire perimeter, about 20 alert armed surveillance aircraft would be needed to defend against bomber-launched cruise missiles; factoring in the assumed 40 percent strip alert rate means that a total force of 50 aircraft would be required.

Similarly, in an attack in which 350 submarine-launched cruise missiles were massed in one area, the 3 alert aircraft that would be needed to destroy the attack could form a continuous barrier of only 720–800 n.mi., which would be only about 20 percent of the total length of the offshore perimeter they might have to defend. The portion of the alert armed surveillance aircraft force defending against submarine-launched cruise missiles would have to be increased to 15 to provide complete coverage. Factoring in the assumption of a 20 percent alert rate for aircraft maintaining continuous airborne alert leads to an inventory requirement of 75 armed surveillance aircraft to defend against submarine-launched cruise missiles. In total, 125 armed surveillance aircraft would be needed.

In making this calculation, we have assumed that the Soviet Union would deploy its submarine-launched cruise missile force in at least 15 submarines. If the Soviets were so accommodating as to deploy their cruise missiles in a smaller number of larger submarines, the United States could adopt an alternative (and cheaper) air defense strategy by taking advantage of its good submarine detection capabilities to station an armed surveillance aircraft above each submarine that approached close to the coasts.

All of the 35 alert armed surveillance aircraft would need a full load of air-to-air missiles. Additional missiles would be required for training and maintenance and to avoid having to move missiles from one aircraft to another whenever an aircraft came off alert. We have assumed that missiles sufficient to arm 70 percent of the total fleet of armed surveillance aircraft would be adequate, which would mean a purchase of approximately 8,750 air-to-air missiles.

Tanker requirements. The effectiveness of the battle management, armed surveillance, and early warning aircraft could be improved by giving them an aerial refueling capability and providing them with tanker aircraft. Currently, about 350 U.S. strategic forces aircraft are supported by about 700 KC–135R equivalent tanker aircraft (assuming that the new KC–10A tanker is equivalent to three KC–135s). Because the various types of aircraft utilized in the Alpha system would be more modern and fuel-efficient than existing strategic aircraft, and because the distances over which the refueling aircraft would have to operate would be much shorter in the Alpha system than they are at present, we have assumed that 1 KC–135R equivalent tanker would be sufficient for every 2 strategic defense aircraft. (At present, 2 tankers are required for every 1 strategic aircraft.) The total tanker requirement in the Alpha system would thus be about 35 KC–10A equivalent aircraft.

Air base and shelter requirements. The aircraft required both for the ballistic missile and the air defense components of the Alpha defense would have to be dispersed across a set of air bases protected sufficiently against ballistic missile attacks to make the price of attacks on these bases every bit as expensive as a direct attack on U.S. strategic forces. Otherwise, a winning strategy would be available to the attacker.

To ensure this relative invulnerability, we first note from Table 1 that approximately 5,100 warheads could be expected to remain on U.S. strategic forces bases at the time a ballistic missile attack might arrive. The ballistic missile defense component of the Alpha system has been structured so that the cost of destroying these warheads directly would be as close as possible to this same 5,100 total.

To extract a price for destroying the defense system by means of attacks on the air bases utilized by the air defense aircraft, we must disperse the defending aircraft across a set of bases that collectively are comparably well defended against ballistic missile attacks. If extracting an appropriately large price was the only consideration, we conceivably could use a single air base defended so well as to extract a price of 5,100 warheads for its destruction. Many factors dictate that a larger number of air bases be used, however.

There is some advantage to be gained, for example, by basing the aircraft near the areas in which their airborne duties would be performed. It also would be advantageous to draw on reserve

military personnel rather than active forces for some of the air operations, which means depending on personnel located in areas in
which other employment is available. Moreover, if too many aircraft
were located on one base, the defense system could be too vulnerable to sabotage. Perhaps most important, crowding too many aircraft on a single air base would make it impossible for the required
numbers to escape quickly in the event of tactical warning of a
ballistic missile attack.

Reflecting these factors, we have elected to use 32 additional
bases for the total of approximately 275 strategic defense aircraft
of the Alpha system. This works out to an average of slightly more
than 8 aircraft per base, which is roughly the base loading employed
by the U.S. strategic bomber force projected in Table 1. We have
assumed that the AWACS and fighter interceptor aircraft already
in the planned U.S. force would be based at ten existing locations
that would become part of the larger base structure. Thus, 32 new
bases would have to be built, or existing but inactive military airfields refurbished. The cost of either option would probably be
about the same.

To extract a total price of 5,100 for the destruction of the 42
strategic defense air bases would require that each base be well
enough defended to extract a price of about 120 attacking RVs for
its destruction. Referring to Table 2, this implies that each base
would need to be defended by 122 HEDI, plus about 20 LEDI,
missiles, and also be assigned 1 ground-based radar. Readers should
note that our analysis assumes that the Soviet Union could not identify a limited subset of the air defense bases whose destruction
would result in the collapse of the entire defense system. This
means that each type of defensive aircraft would have to be spread
across the available bases as uniformly as possible.

The Alpha strategic defense system would require one final construction element. Because we have assumed that atmospheric filtering would be used to discriminate between RVs and lightweight
decoys, which would entail a defensive battle within the atmosphere, and because we have elected to fire LEDI missiles only
if the HEDI missiles assigned to any given warhead missed, the
minimum range from the defended target at which the LEDI missile
would intercept the attacker could be very low, perhaps as low as
10,000–15,000 ft. If an attacking warhead was fuzed to detonate

at this distance, or was fuzed to explode instantaneously upon sensing that it had been intercepted, the nuclear effects felt by the target could be sufficiently severe to destroy such relatively ''soft'' objects as normal hangars and aircraft. Thus, it would be difficult to extract a significant attack price unless the targets (U.S. strategic forces) were hardened reasonably well against the effects of nuclear explosions.

This hardening presents little problem for land-based missiles, which would be located in blast-resistant silos or hardened mobile launchers. A successful defense of bomber bases, however, would require that the bases be hardened sufficiently to survive a nuclear detonation at a distance at which the second intercept attempts would take place. The cost of building concrete shelters for strategic bombers at the 30 bases assumed to be used by these aircraft are thus included in the cost of the Alpha system. Also, concrete shelters would have to be built for the several types of missile and air defense aircraft included in the system; the costs of such shelters at the 10 air bases already used by air defense interceptors and at the 32 additional bases that would have to be constructed are also included. (Shelters obviously would not have to be provided for alert aircraft.)

COSTS OF THE ALPHA COMPONENTS

We are now in a position to estimate the cost of each component of the Alpha system. Table 4 summarizes the numbers of the various elements that would be required to make it unprofitable for an enemy to attack the bases of U.S. strategic forces. The cost of each element is described below.

High Endo-Atmospheric Interceptors

Intercepts in the upper layer would be carried out by high-acceleration missiles that would meet incoming RVs at an altitude of at least 100,000 ft. We have assumed that atmospheric drag would have had sufficient discriminatory effects to permit the laser radar/battle management aircraft to distinguish the real reentry vehicles from accompanying decoys at about 270,000 ft. The laser

TABLE 4
Alpha System Components

COMPONENTS DEPLOYED AT STRATEGIC FORCES BASES

Type of Strategic Force	Number of Bases	Total Number of HEDI	Total Number of LEDI	Total Number of Radars
Trident Strategic Submarines	2	1,152	230	2
Bombers	30	1,800	300	30
MX	50	500	100	5
Minuteman III	550	1,650	0	0
Midgetman	3	51	9	3
Air Bases for Defensive Aircraft	42	5,124	840	42
Total	677	10,277	1,479	82

UNDIFFERENTIATED COMPONENTS

Element	Number
Laser Radar/Battle Management Aircraft	40
Early Warning Air Defense Aircraft	75
Armed Surveillance Aircraft	125
Aerial Refueling Tankers	35
Air-to-Air Missiles	8,750
Air Base (additional)	32
Aircraft Shelters	435

NOTES: Excepting 10 of the 42 air bases for defensive aircraft, these figures include only those components of the strategic defense system Alpha that are not already included in the U.S. defense program. Components of the system that already exist or are planned include early warning satellites, over-the-horizon radars, 300 interceptor aircraft, 10 airborne warning and control systems, 10 bases for air defense aircraft, and existing bases for strategic submarines, land-based missiles, and strategic bombers.

radar/battle management aircraft would direct the missile toward its target; the missile would then detect the target RV, home on it with on-board sensors, and hit to kill.

A new missile would have to be developed for this purpose, although all necessary technologies are well in hand. Manufacturers would have substantial opportunity to compete to produce the missile more cheaply because it would be produced in such large quantities. Informal discussions with knowledgeable individuals suggest that $2 billion is a reasonable estimate of the interceptor's development cost. (Readers should note that unless stated otherwise all costs presented in this book are given in constant 1987 prices.)

The same sources suggest that it might cost about $6 million to produce the first HEDI interceptor.[12] Given the large acquisition requirements, however, we can assume that a "learning curve" would reduce the marginal cost of a single missile by 10 percent for each doubling of the total number of missiles to be produced. Thus, the second interceptor would cost $5.4 million, the fourth $4.9 million, and so on. Although such a learning curve would be better than has been true of most defense systems produced in large quantities, it is not unreasonable to expect. The sensitivity of our overall results to this assumption is demonstrated in chapter 6.

We have assumed that taking maintenance, testing, training firings, and other operational needs into account, 80 percent of the missiles produced could be kept on-line; this is also an optimistic assumption based on past experience. Calculation of the cost of producing the necessary 12,900 high endo-atmospheric interceptors then becomes straightforward; it would come to a total of about $21 billion, or about $1.6 million per missile.

In our Alpha system the HEDI missiles would be based in groups of ten, with the groups being separated by sufficient distances so that the destruction of each group would require a separate nuclear weapon. The groups would be located on military bases whenever possible. In all cases, they would be fenced and instrumented to prevent intrusions, but they would be unmanned. The development cost for the basing concept is likely to be around $200 million and the cost of acquiring each such missile base, including necessary communications and security systems, on average, perhaps $3 million. Total basing costs for the upper layer interceptors would then be roughly $3.1 billion.

The operating costs of this missile force should be relatively low, primarily because of its very limited requirements for manpower. They would consist of little more than the costs of maintaining the missiles, their bases, and the ground-based portions of associated communications systems linking the missiles to their controlling battle management aircraft. We have assumed, conservatively, an annual operating cost of 5 percent of the procurement cost of the missiles and their bases. Thus, the total annual operating cost of the complete force of upper layer interceptors would be approximately $1.2 billion.

Low Endo-Atmospheric Interceptors

The lower layer defense system would consist of 1,479 LEDI missiles controlled by 82 ground-based radar installations. These interceptors would have to be very high-acceleration missiles in order to intercept their targets in the 8 to 10 seconds between an observed miss by HEDI missiles and the attacker's arrival at a point close enough to destroy the target. We have assumed that the low endo-atmospheric missiles would employ semi-active radar homing rather than the more sophisticated seekers employed by the upper layer missiles. We have further assumed that the LEDI missiles would hit to kill.

We estimate development costs of roughly $3 billion. Given the very high acceleration required, we have assumed a first unit production cost of $9 million.[13] As with the high endo-atmospheric interceptor force, we have assumed that 80 percent of the missiles produced could be kept on-line and that a 90 percent learning curve could be achieved, leading to total production costs of approximately $5.9 billion, or approximately $3.2 million per missile.

We have also assumed that LEDI missiles, except for those guarding Midgetman bases, would be based in groups of roughly ten. Per group, basing and operating costs should be roughly the same as those for the HEDI missiles, for a total basing cost of about $450 million and annual operating costs around $320 million.

Ground-Based Radars

We have assumed that the ground-based radars that would control the lower layer missile defenses would be a substantially upgraded version of an existing surveillance radar such as the TPS–43E. Development of an improved version would cost about $500 million and procurement of the relatively small number of radars needed would cost about $20 million each.[14] With a small operating crew at each site, these radars should cost about $3.5 million to operate at each site every year. Summing these unit costs leads to a total cost of about $1.6 billion to procure and install the 82 radars needed and to a total annual expense of about $290 million to operate the radar system thereafter.

Laser Radar/Battle Management Aircraft

We have assumed that modified KC–10 cargo aircraft would constitute the laser radar/battle management aircraft controlling intercepts in the upper layer of defense. Based on informal discussions with manufacturers, these aircraft and the complex suite of equipment they would carry would cost $4 billion to develop, $250 million each to procure (including the laser radar and other battle management equipment), and $15 million per year to operate per aircraft. Once developed, the required fleet of 40 aircraft would thus cost $10 billion to procure and about $600 million per year to operate.

Armed Surveillance Aircraft

The armed surveillance aircraft would be modified KC–10 type aircraft with conformal radars. We estimate that the aircraft and the required fire control and munitions handling systems they would carry would cost roughly $3 billion to develop, $150 million each to procure, and about $9 million per aircraft to operate each year. Once developed, the required force of 125 such aircraft would cost about $19 billion to procure and about $1.1 billion per year to operate.

Air-to-Air Missiles

Based on informal discussions with manufacturers, and an examination of cost data for such comparable missiles as the Phoenix and AMRAAM, we have assumed a development cost of $1 billion for the air-to-air missile, a first union production cost of $4 million, and a 90 percent learning curve. These assumptions yield a total procurement cost of about $10 billion for the necessary production run of 8,750 missiles. Assuming, as we have above, that the annual operating cost of a missile system is roughly 5 percent of its procurement cost, total annual operating costs for this element of the Alpha system would come to about $500 million.

Early Warning Aircraft

We estimate $500 million for developing the required modifications of the TR-1 type aircraft that would be employed for an early warning role, and about $50 million for purchase of each aircraft with the required radar.[15] We have assumed that the aircraft could be operated for about the same cost as an F-16—approximately $2.5 million per year per aircraft. Once developed, the required force of 75 such aircraft would thus cost about $3.8 billion to purchase and about $190 million per year to operate.

Tankers

KC-10A tankers cost about $135 million each.[16] Because the aircraft is simpler than the laser radar/battle management variant described previously, and because it would require a smaller maintenance and flight crew (most likely of reservists), we have assumed that it would cost $7 million per aircraft per year to operate. The tanker force of 35 aircraft would thus cost about $4.7 billion to procure and $250 million per year to operate.

Air Bases

According to informal U.S. Air Force estimates, the cost of building an austere air base is between $50 and $60 million. We have used the lower figure on the assumption that some of the air bases would undoubtedly be created by reopening and refurbishing existing airfields no longer in use. The air force also estimates that the cost of operating an austere air base is about $40 million per year.[17] The 32 additional air bases required to support the several types of strategic defense aircraft utilized in the Alpha system would therefore cost about $1.6 billion to build and about $1.3 billion per year to operate.

Hardened Aircraft Shelters

The aircraft shelters now being built in Europe for NATO tactical aircraft can hold several fighters each; each costs around $1.2 million.[18] Because it would be both necessary to build larger shelters to house strategic defense aircraft and disproportionately more difficult to harden them, we have assumed that each shelter would cost around $2 million. In all, 435 shelters would be needed for all non-alert aircraft included in the Alpha system, at a total cost of about $0.9 billion.

Other Costs

The costs of individual components of the Alpha system do not take into account the significant costs of some additional requirements: (1) considerable augmentation of the North American Air Defense Command (NORAD) to accomplish this much expanded mission, (2) research, development, and procurement of the command and control systems needed to bind together individual elements of the system, and (3) additional investments in the logistics support system, such as maintenance and support facilities for the new types of missiles and aircraft. Such costs are difficult to assess in specific terms. Therefore, we have included an allowance of $2 billion in procurement costs and annual operating expenses of $200 million to cover them.

Total Alpha System Costs

Table 5 summarizes the individual system component costs described above. The bottom line suggests that it will cost approximately $160 billion to develop and procure all components of the Alpha strategic defense system and to operate them for ten years.[19] Only four components would account for approximately 70 percent of this total: the high endo-atmospheric missiles used in the upper ballistic missile defense layer, the laser radar/battle management aircraft used to control the upper layer of ballistic missile defenses, and the armed surveillance aircraft together with their air-to-air missiles.

Readers should note that the Alpha system component intended to defend against bombers and cruise missiles combined with those

TABLE 5
Costs of Strategic Defense System Alpha
(billions of 1987 dollars)

Component	R&D	Procurement	Construction	Ten-Year Operations	Total Cost
HEDI	2.0	21.0	3.1	12.0	38.1
LEDI	3.0	5.9	0.5	3.2	12.6
Ground Radars	0.5	1.6	—	2.9	5.0
Battle Management Aircraft	4.0	10.0	—	6.0	20.0
Armed Surveillance Aircraft	3.0	18.8	—	11.0	32.8
Air-to-Air Missiles	1.0	10.0	—	5.0	16.0
Early Warning Aircraft	0.5	3.8	—	1.9	6.2
Tankers	—	4.7	—	2.5	7.2
Air Bases	—	—	1.6	13.0	14.6
Aircraft Shelters	—	—	0.9	—	0.9
Other Costs	—	2.0	—	2.0	4.0
Total	14.0	77.8	6.1	59.5	157.4
Rounded Total	**10**	**80**	**10**	**60**	**160**

NOTES: The degree of precision expressed by these figures is not meant to suggest a comparable degree of certainty about the estimates. It is necessary in cumulating the costs of the system to keep track of the individual elements at this level. The final "rounded" line is the more appropriate way to think of these estimates.

portions of the ballistic missile defense component that would guard the bases used by air defense aircraft constitute nearly two-thirds of the total cost of the complete system. A decision to defend U.S. strategic forces from only ballistic missiles would reduce the ten-year systems cost of the Alpha system to perhaps $40 to $50 billion, depending upon the assumptions made about learning curves and other economies of scale. Such a decision, however, would leave open strategic options that the Soviet Union could exercise to defeat the assumed purposes of deploying this type of strategic defenses.

Moreover, if the United States chose to defend only its land-based ballistic missiles (as listed in Table 1) and to defend them from attacks by ballistic missiles only—an even more modest strategic objective sometimes suggested—and applied the same techniques considered in the Alpha system, the total ten-year cost for such a system would be roughly $30 billion. This figure includes nearly $10 billion in development costs for laser radar/battle management aircraft and the HEDI and LEDI missiles. Defense of the land-based ballistic missile force alone might be less expensive if a single layer defense was employed using only ground-based radars.

2.
NOTIONAL STRATEGIC DEFENSE SYSTEM BETA
Defense of Nuclear Retaliatory Forces plus a Partial Population Defense Against Small Attacks

T HE SECOND STRATEGIC DEFENSE SYSTEM would build upon the first, adding the components necessary to defend the most densely populated urban areas of the United States and Canada against relatively small ballistic missile attacks. Like the Alpha system, the Beta system would utilize only weapons, sensors, and command and control systems that could be built today or in the near future; it could be fully deployed very early in the next century.

In terms of cost, it would be relatively inexpensive to add a light area defense capability to the Alpha system, as most of the necessary components would already have been put in place in order to defend nuclear retaliatory forces. Perhaps most important, as the Alpha air defense component would have been sized to defend against the entire projected force of Soviet strategic aircraft, the Beta system would require no additional air defenses.

In operational terms, the purpose of the light area defense component of the Beta system would be to raise the required size of any attack intended to destroy the more densely populated portions of the United States. Consistent with the required extension to southern Canada of the Alpha air defense component, the Beta system also would include a light ballistic missile defense capability for Canada's most densely populated areas.

This area defense capability of the Beta system would provide protection against accidental launches of small numbers of nuclear-armed missiles. It also could reduce significantly the total damage that could be inflicted on the United States and Canada by countries with small nuclear forces. In addition, the Beta system also could make limited "demonstration" attacks by the Soviet Union or another major nuclear power somewhat more difficult. These potential achievements could become more important in the next century as additional nations acquire nuclear-armed ballistic missiles.

The Beta system's light area defense component would not provide protection against any major attack; even a relatively small nuclear power could overcome the system by concentrating its missiles against a small number of the most important urban targets or by attacking larger numbers of undefended targets. However, the Beta system could be used as the "last ditch" terminal layer of a more extensive ballistic missile defense system that included components to destroy attacking missiles in space. In such a role the Beta system would seek to destroy those few ballistic missile reentry vehicles that had leaked through the previous defensive layers. The Gamma and Delta systems described later in this book incorporate elements of the Beta system for this purpose; only such a comprehensive system potentially would be able to achieve the most ambitious objectives conceivable for strategic defenses.

SIZE AND CHARACTERISTICS OF THE ADDITIONAL ELEMENTS NECESSARY TO EXPAND FROM THE ALPHA TO THE BETA SYSTEM

Unlike defending nuclear-armed delivery systems, described in the Alpha system, there is no practical means of "hardening" cities so that they could survive nuclear detonations at short ranges. A nominal 1-megaton nuclear weapon could cause severe damage to residential housing at ranges of 5–8 miles when detonated at its optimal height of burst; light damage would be caused at ranges out to 13 miles.[20]

We have assumed that the goal of the light area defense would be to prevent damage from nominal 1-megaton yield attackers, which would be greater than light damage to residential housing. This implies that the attackers would have to be prevented from getting closer than approximately 65,000 ft. With the high endo-atmospheric interceptors meeting their targets at an altitude of roughly 100,000 ft., leakers would have to come down only 35,000 ft. further before they could defeat the defender's goal. This equates to between 3 and 5 seconds of travel time for the attacker, which is far too short a period for the launch and fly-out of a low endo-atmospheric interceptor.

We could, of course, set a more modest goal for low endo-atmospheric interceptors in the light area defense role. For example, they could be used to prevent leakers from causing severe damage to reinforced concrete buildings—a goal they could meet by keeping the nominal 1-megaton yield attacker from getting closer than roughly 17,000 ft. In view of the limited benefit of accomplishing such an objective, however, it would seem more reasonable to invest in a stronger high endo-atmospheric defense instead.

Determining the number of interceptors necessary for a light area defense is fairly arbitrary. After all, what does "light" mean? Does it mean that interceptors should be so spread across the United States as to allow at least 1 interceptor to reach the first reentry vehicle aimed at any point in the country? Such a strategy would value the protection of all areas equally, regardless of the number of people or any other items of special national value that might be located in any particular area.

It would seem more reasonable to allow a distribution of interceptors across areas that would be proportional to the values located in that area. At the same time, certain threshold costs are associated with setting up defenses in any area. Thus, for example, interceptors probably should not be located in any area whose value was so low as to merit only 1 or 2 of them.

In view of these considerations, we arbitrarily have assumed that HEDI interceptors intended for area defense would be deployed in bases of 10 missiles each and that 1 base would be assigned for every 500,000 people located in the forty largest metropolitan areas in the United States. These areas contain approximately 110 million people, which is somewhat less than one-half of the U.S. population. The Canadian portion of the Beta system was sized in a similar, but slightly different fashion: we provided 1 base of 10 interceptors for every 500,000 people within those Canadian metropolitan areas with total populations of over 500,000. About 40 percent of the Canadian population is located in seven metropolitan areas containing nearly 10 million people and would qualify for protection under this criterion.

In sizing the Beta defense, we further assumed that the defenses associated with strategic bases would be coordinated with those protecting nearby cities. Plotting the locations of current strategic forces bases and air bases that would be likely locations of the strategic air defense components of the Alpha system shows that nearly one-half of the largest U.S. metropolitan areas would be within 20 miles of a defended base. We assume, however, that the allocations of defenses to large metropolitan areas and nearby bases would be made independently.

Readers should note that under this criterion the largest cities would receive very large allocations of HEDI missiles. The New York metropolitan area, for example, would be allocated roughly 36 interceptor bases with 360 missiles. Since the interceptors employed could protect areas from light residential damage as far as 40 miles away, it is clear that the missiles protecting New York would have to be carefully coordinated and that tapered salvo firing doctrines similar to those employed for the low endo-atmospheric interceptors of the Alpha system also should be employed for the light area defense component of the Beta system.

Given the above assumptions, the light area defense component of the Beta system would require 220 missile bases containing 2,200 high endo-atmospheric interceptors—about one-fifth the number of HEDI missiles needed for the Alpha system. This number is sufficiently small that the number of laser radar/battle management aircraft used to control outer layer intercepts in the Alpha system would probably not have to be increased. The 40 such aircraft already purchased for that system also could manage the light area defense task, with only modest modifications of the battle management system to allow coordination of the area defenses and those of nearby strategic bases.

COSTS OF THE ADDITIONAL COMPONENTS REQUIRED FOR THE BETA SYSTEM

The only additional costs of the Beta system, then, would be those required to buy, deploy, and operate the additional HEDI missile groups and to modify the battle management system. Using the same assumptions made in the Alpha system estimates, we estimate that a total of roughly 2,750 additional missiles would have to be purchased along with the 220 bases at which the missile groups would be deployed.

It would cost roughly $3 billion to buy 2,750 additional HEDI missiles. Considering that the higher price of land in urban areas probably would triple the assumed cost of each area defense interceptor group base, on average, as compared to the cost of those interceptor bases used to defend missile sites and bomber bases, then the total cost to construct bases would be $2 billion. Our estimate of the operating costs of the Beta system's additional missiles and bases comes to $250 million per year, using the same assumptions as in the Alpha system. We have also allowed $500 million to develop the modifications of the battle management system required to accommodate coordinated defenses of large areas and defenses of nearby strategic bases. There should be no other significant incremental development costs.

TABLE 6
Costs of Strategic Defense System Beta
(billions of 1987 dollars)

Element	R&D	Procurement	Construction	Ten-Year Operations	Total Cost
HEDI	—	3.0	2.0	2.5	7.5
Battle Management Modifications	0.5	—	—	—	0.5
Subtotal	0.5	3.0	2.0	2.5	8.0
Alpha System	14.0	77.8	6.1	59.5	157.4
Total	14.5	80.8	8.1	62.0	165.4
Rounded Total	**10**	**80**	**10**	**60**	**170**

Thus, the incremental ten-year costs of the light area defense element of the Beta system (summarized in Table 6) would come to approximately $8 billion. This is clearly a very modest expense, but it should be remembered that the cost is so low only because the Beta system would use many elements already acquired for the Alpha system; a "stand-alone" light area defense system would be far more expensive. The total ten-year cost of the Beta system equals that of the Alpha system plus the $8 billion increment, or about $170 billion in round numbers. (Costs of a Beta system reconfigured to complement the space-based defense component of the Gamma system are included in Table 8.)

3.
NOTIONAL STRATEGIC
DEFENSE SYSTEM GAMMA
A Comprehensive Defense Using
Space-Based Missile Interceptors

W HILE A LIGHT AREA DEFENSE SYSTEM would be a step toward President Reagan's goal of providing comprehensive protection against nuclear attack, it would not be a very large step. Consisting of only a single layer of less-than-perfect defenses, any light area defense system would inevitably be "leaky" and relatively easily overwhelmed by attacking forces of even modest size. More effective area defenses would require the deployment of additional layers, each of which could compound the others' individual capabilities.

We have assumed that it would prove impractical to intercept reentry vehicles with space-based defensive systems during the period from when they were deployed by the postboost vehicle to when they began to reenter the atmosphere—the midcourse phase of their trajectory. This is because it would be extremely difficult to discriminate between the real warheads and the large numbers of decoys that could be deployed by each missile. The low weight required for decoys not intended to reenter the atmosphere would mean that each Soviet missile could deploy such large numbers of confusing objects that the required number of intercepts in the midcourse phase would be too large for any midcourse defensive system to handle. This assumption is discussed further in chapter 6.[21]

For these reasons, the most attractive means of adding layers to a strategic defense would be to place defensive weapon systems in space, where they could attack enemy ballistic missiles in the earliest phases of their trajectories—the boost and postboost phases. Ballistic missiles are particularly vulnerable during the initial portions of their flights: they move relatively slowly as they accelerate through the atmosphere; they are still dependent on the correct functioning of rocket motors and guidance systems; they still contain tons of highly explosive fuel; they emit bright infrared signatures and thus are easily tracked; they have not yet deployed their multiple warheads and decoys on separate paths.

Thus, both the Gamma and Delta defense systems would add to the Beta system a space-based defense layer to attack ballistic missiles in the earliest phases of their trajectories. The Gamma system would make use of technologies that, for the most part, are close at hand; in particular, space-based interceptor missiles. This system could probably be fully deployed around the year 2012. The Delta system would require more advanced technologies— space-based directed energy weapons. The Delta system probably could not be fully deployed until around 2020.

The primary operational difference between the Gamma and Delta defense systems and their predecessors is that their purpose would no longer be limited. They would no longer seek only to stop small attacks, or to extract a price for a successful attack, or simply to reduce the number of targets an enemy might hope to attack successfully. Instead, the purpose of the Gamma and Delta systems would be comprehensive: to come as close as possible to stopping totally any nuclear attack that would be within any enemy's capacity.

To design a defense system with such a comprehensive objective, we first must make some assumptions as to the likely size and characteristics of the entire Soviet force of long- and intermediate-range missiles at the time such a system would be functioning. Within broad limits, these assumptions are arbitrary.

Soviet representatives have stated that in the event that the United States deployed ballistic missile defenses, the USSR would expand and improve its ballistic missile forces sufficiently to overwhelm whatever system the United States had attempted to establish. Whether or not they could do this would depend on a plethora of factors far too complex to delve into in this book. It is possible, at the same time, that (current Soviet statements notwithstanding) the Soviet Union could decide later in the century, either tacitly or as the result of negotiations, to follow the U.S. lead and concentrate on deploying effective defenses of its own. In such a case the USSR might not improve its offensive forces and might even agree to mutual reductions in offensive forces. Such a mutual transition to defensively oriented postures could result in a more stable relationship between the United States and USSR. The forty-year history of the strategic competition between the two nations suggests that such an outcome would be unlikely, however, unless the two nations departed sharply from their past patterns of behavior.

In view of this uncertainty we have assumed that a Soviet long-range ballistic missile (ICBM) force of essentially the same size as the current force would consist of 1,400 land-based missiles equipped with an average of 8 warheads each, a submarine-launched ballistic missile (SLBM) force of 1,000 missiles averaging 4 warheads each, and a force of intermediate-range, land-based ballistic missiles (IRBMs) of 600 missiles averaging 3 warheads each.[22]

There is no special reason to believe that the Soviet missile force in the year 2010 would be essentially the same size as it is today, but neither is there any decisive reason to postulate that the force would be either significantly larger or smaller. It does seem unreasonable to assume, however, that the characteristics of Soviet missiles will not be changed, thereby complicating the problems of defense. There probably would be fifteen to twenty years between a U.S. commitment to full-scale development of a comprehensive defense system and achievement of that system's operational status—more than enough time for the USSR to deploy a new generation of offensive missiles designed, in part, to capitalize on the vulnerabilities of whatever defense system the United States had selected. Indeed, it would be imprudent for the United States not to assume such a Soviet reaction and not to design a defensive system to minimize such potential vulnerabilities.

Consequently, we have assumed that the Soviet missile force to be deployed in the early part of the next century would show three important changes (a fourth change will be described in connection with the Delta system):

(1) The amount of time that Soviet missiles spend in the boost phase would have been reduced to near 90 seconds, as compared

to the 150 to 300 seconds that the Office of Technology Assessment (OTA) estimates for current and emerging Soviet missiles.[23] Such a reduction, which would more than halve the time during which Soviet missiles in the boost phase would be vulnerable to attack, would be relatively easy to accomplish with available technologies. Boost phases as short as 50 seconds are considered feasible by many experts.

(2) To complement this shortening of the boost phase, the Soviet Union would shorten the postboost phase of flight for its missiles equipped with multiple RVs—the phase during which a missile maneuvers to set each of its reentry vehicles on its separate trajectory and dispenses associated decoys. We have assumed that this phase, which now takes several minutes, would be reduced to 60 seconds. While even shorter times are conceivable, they could lead to a combination of difficulties, including a large reduction in the maximum distances between the targets that could be attacked by RVs from the same missile, reductions in accuracy, a requirement for separate thrusters for each reentry vehicle and its associated decoys, difficulties in dispensing decoys, and/or unacceptably high costs to minimize these problems. Reduction of the postboost phase to even 60 seconds would increase Soviet missile costs in any case.

(3) The Soviets would compress the area in which their land-based missiles are deployed by two-thirds.[24] As will be described more completely below, such a concentration of the Soviet ICBM and IRBM force would greatly increase the required density of U.S. defensive weapons deployed in space and thus the potential cost of any defense system incorporating a space-based, boost-phase layer. This change also would impose significant extra costs on the Soviet Union because of the higher expenses associated with mobile missiles or the construction of new silos.

There is of course no way of knowing how, precisely, the USSR would respond to a U.S. decision to develop and deploy a strategic defense system. The three steps just described are well within Soviet capabilities, however. We discuss the ramifications of the three assumptions separately below. The sensitivity of our overall cost estimates to changes in the size or characteristics of the Soviet missile force is described in chapter 6.

Size and Characteristics of the Additional Elements Necessary to Transform the Beta System into the Gamma System

The space-based component of the Gamma system would consist of four elements: (1) interceptor rockets, each of which, upon detection of the launch of attacking missiles, could be assigned to an individual target and directed to an intercept point in space from which to home on the target and destroy it by direct contact; (2) battle satellites in which the interceptor rockets would be clustered, maintained in low earth orbits, along with decoy battle satellites to complicate any attempt to attack the defense system; (3) a battle management system, which would be located in satellites in 5,000-km. orbits; and (4) a space launch system to deploy and maintain the several types of satellites.

Interceptor Rockets

The space-based interceptors in a Gamma defense would be launched as soon as the launch of Soviet missiles had been detected and confirmed. We have assumed that detection would be accomplished immediately upon launch by relatively simple space-based radars that would be designed solely for this purpose. We have further assumed that confirmation of a Soviet missile launch and a decision to launch the required U.S. space-based interceptors could be accomplished within 10 seconds.

The interceptor would accelerate rapidly toward an intercept point defined by the battle management system. Midcourse corrections would be calculated by the battle management system and relayed to each interceptor as it moved toward its target. As each interceptor approached its target, an on-board guidance system, similar to that incorporated in the U.S. "Space Defense System" now being developed as an antisatellite weapon, would cause the interceptor to collide at a very high speed with the target missile and destroy it.

We have assumed that interceptors would be capable of achieving a single-shot probability of kill of 0.9 against any Soviet missile within range. If each target was exposed to only one intercept attempt, the kill probability, although high, would not be high enough to achieve the kind of overall effectiveness expected of the Gamma system. Consequently, we have assumed that additional interceptors, 10 percent as many as constituted the first wave, would be timed to arrive in the vicinity of the offensive missiles about 10 seconds later. The interceptors in this second wave would be redirected by the battle management system to those attacking Soviet missiles that had been observed to survive the first wave of attacks.

We further have assumed that the battle management system would be capable of making a nearly perfect overall assignment of interceptors to target missiles, thus avoiding the twin potential errors of sending more interceptors than necessary to kill some Soviet missiles while failing to assign any interceptors to others. These battle management procedures, as well as those involved in coordinating the first wave of interceptors, would represent a substantial engineering challenge. Assuming such capabilities could be realized, a space-based system of the type described would have a theoretical leakage rate of 1 percent—assuming of course that there were enough interceptors to attack all the missiles that the Soviet Union could launch.

To determine how many interceptors would be needed to attack all the Soviet missiles, we must first determine how long a period would be available for an interceptor to fly from its initial position in space to its point of intercept with the target missile. To establish a reasonable value for this length of time, we must examine its effects on the design of the space-based component of the Gamma system.

During the period that the Soviet missile engine was firing, a bright infrared signature would be available for detecting and tracking the missile. If the offensive missile shut down its motor and coasted for a period, the interceptor could still meet it by flying toward a projected ballistic intercept point and then homing on the target. Given the physical size of the target and the contrast of its infrared signature with even an earth background, the offensive missile should be sufficiently detectable for such defensive actions. The postboost vehicle that a "MIRVed" missile would use to maneuver each reentry vehicle and its associated decoys to their separate trajectories also should be readily detectable. We have assumed that the defending interceptors would be able to pick out the correct object to strike from among the various diversions that could be present, such as the booster's plume or the booster itself after it has separated from the postboost vehicle.

The best strategy for an offensive missile would be to complete its main boost as rapidly as possible and run quickly through the postboost maneuvers necessary to dispense the payload. The best strategy for the interceptor would be to complete its attack before the attacking missile could begin to dispense its payload. The length of time from detection of the offensive missile launch until when the missile began dispensing its payload would thus become a critical parameter for both sides. The longer this period, the more time available for the defense to bring interceptors from distant battle satellites to bear on the target. The more distant the points from which interceptors could be fired by the defense, the greater the fraction of the total interceptor force that could be brought to bear and, correspondingly, the lower the total number of interceptors required in space.

As noted, the boost-phase portions of the trajectories of current Soviet long-range ballistic missiles vary between two and one-half and five minutes. Shorter-range submarine-launched ballistic missiles are likely to have shorter boost phases, and such intermediate-range theater nuclear missiles as the SS–20 should have still shorter boost times yet. We should expect future generations of Soviet ballistic missiles to be designed with shortened boost phases in mind, as this will impose a large increase in the cost of U.S. space-based defenses.

In these calculations we have made the conservative assumption that the defense would prove effective against all Soviet long- and intermediate-range missiles if it employed an interceptor missile capable of covering the entire intercept trajectory in 100 seconds. Taking into account the confirmation and decision delays of the system, we are thus saying that an interceptor could be considered effective if it hit the missile booster, or the postboost vehicle, within 20 seconds after shutdown of the booster—when most of the reentry vehicles would likely still be on board.[25] (In chapter 6 we provide an indication of how sensitive the Gamma system's total cost would be to variations in this maximum allowable interceptor flight time.)

We have assumed that an interceptor missile carrying a 5-kg. homing warhead and having an initial weight of 150 kg. could be built to achieve a burn-out velocity of 10 km./sec. (Many experts consider it optimistic to assume that as complex a warhead as the one needed here could be built at so light a weight.) The first wave of missiles would keep a minimum reserve of thrust capability to cope with uncertainties in where the target was headed. They thus would accelerate to perhaps 9 km./sec. almost immediately and average slightly more than this speed during the flight. We have assumed that the homing warhead would have a total maneuver capability of perhaps 0.5 km./sec. The second wave of interceptors would accelerate immediately to 8 km./sec., holding the remaining acceleration capability in reserve to allow redirection of the interceptor to a missed target.

The range of interceptors in the second wave would establish the limit on the effective coverage range of the space-based interceptors, were it not for the fact that, because of the potential vulnerability of the battle satellites, interceptors would presumably be clustered in groups substantially smaller than the maximum allowed by their effective range. The extra battle satellites thus deployed to reduce vulnerability would create a spatial distribution of interceptors, which would guarantee that the group assigned to attack the Soviet missiles rising from any particular area would include some interceptors within considerably less than maximum range of their targets. These latter interceptors would be the ones assigned to the second wave of defenders. This consideration makes the range capability of the interceptors in the first wave the dominant factor in determining effective coverage range.

Assuming a 35g acceleration for the interceptors, those in the first wave could reach their 9 km./sec. initial fly-out speed in 26 seconds. During this acceleration phase the interceptors would cover a distance of nearly 120 km. During their remaining maximum allowable flight time of 74 seconds, they could cover 666 km., for a total of about 780 km. We have assumed that the interceptors would be located initially at orbital altitudes of 400 km. and that the attacking Soviet missiles would be at altitudes of approximately 100 km. at the time of intercept. These parameters imply that the interceptors could reach Soviet targets inside a circular area at the intercept altitude with a radius of 720 km. or, equivalently, that the space-based defense component must be configured to have 1.1 ready interceptors requiring less than a 720-km. horizontal displacement from orbit in order to hit each Soviet missile.

This requirement points to the costs that the Soviets could impose on the defense by reducing their missile deployment area. If, for example, all 2,000 of the assumed Soviet land-based missiles were located in an area of 100 sq.km., the required density of space-based interceptor missiles would be about 2,200 for an area in space of about 1.6 million sq.km. This required density of in-range interceptors would lead to a total requirement of roughly 710,000 ready interceptors in orbit.

Reducing the missile deployment area could have significant drawbacks for the Soviet Union, however. Unless the Soviets had deployed reliable means of protecting the locations of the offensive missiles, these sites could be destroyed in a preemptive attack before the missiles were launched. (Concentrating the missiles in this manner would make the task of defending them against ballistic missile attacks somewhat easier.) Moreover, dispersal of missiles across a wide area would significantly extend the area they could threaten. These considerations would be particularly important in the case of submarine-launched ballistic missiles.

In these calculations we have assumed that the Soviet Union would disperse its land-based missiles roughly uniformly across an east-to-west swath of its territory comparable in its 6,000-km. length to that used now, with an average depth of 500 km. in its north-to-south dimension. As noted previously, this area would equal roughly one-third of the land area over which Soviet land-based, long-range missiles are deployed currently. Such a compression could be achieved readily, however, as new generations of Soviet missiles are deployed.

We have assumed that Soviet submarine-launched ballistic missiles would be distributed roughly uniformly across the 8-million-sq.km. area consisting of the Sea of Okhotsk and the portions of the Arctic Ocean between the Soviet Union and the North Pole. A space-based defense system that employed polar orbits to allow coverage of areas near the poles, but that was sized to deal with the USSR's land-based missiles, would lead to a system with very high concentrations of space-based interceptors at the higher latitudes. Consequently, concentration of Soviet submarines within a small portion of this potential Arctic deployment area should not pose a particular problem to the defense, assuming that deployment was not carried to an extreme. It is reasonable to assume that the Soviets would not do so, since concentrating the submarines would likely increase their potential vulnerability. The USSR could deploy its submarines in additional ocean areas, of course, but it currently places a premium on keeping most of its strategic submarine force reasonably near their home ports, so that they can be more readily protected from U.S. antisubmarine warfare forces. We have assumed that this practice would not change.

Finally, we have assumed that the Soviet Union would be unable to implement strategies that sought to exhaust the U.S. space-based defenses by launching appropriately chosen patterns of their land-based missiles nearly simultaneously and then launching their submarine-based missiles at a time calculated precisely to challenge the depleted portions of the U.S. defense satellite constellation. We have assumed that the United States would be capable of maneuvering its defense satellites to cover such weaknesses in a sufficiently timely manner. Here again, we have made an assumption that is optimistic for the defense.

Given these assumptions, the land-based missile force would present the higher concentration of attacking missiles to the defense and thus would determine the necessary density of interceptors in space. That density is equal roughly to 1.1 times the number of Soviet land-based missiles, divided by the size of an area above the surface of the earth that would be within approximately 720 km. of the projected Soviet missile deployment area.

Considering only Soviet long-range, land-based missiles (ICBMs), 1 interceptor would then be required for every 9,370 sq.km. of the entire surface area of a sphere concentric with the earth and of a radius equal to the earth's plus the orbital altitude of the interceptors. If intermediate-range, land-based missiles (IRBMs) are also considered, then 1 interceptor would be required for every 6,560 sq.km. of the surface area of such a sphere. The total number of ready interceptors needed in space can now be estimated by dividing the surface area of this sphere, which is about 570 million sq.km., by 9,370 or 6,560, respectively. The number of required ready interceptors in space thus comes to about 60,800 for the ICBM case, and 86,900 if intermediate-range missiles are considered as well. As noted above, either system also should be able to defend against SLBMs launched from the Arctic Ocean or the Sea of Okhotsk.

Battle Satellites

The interceptor missiles would not be placed in space singly, of course, but clustered on satellites that would maintain the missiles in ready-to-fire condition, report on their condition, receive initial firing instructions from the battle management satellites, and launch interceptors when necessary. The economies of scale of these support operations suggest that these "battle" satellites should collect together as many missiles as possible.

The larger the number of interceptors based in a single satellite, however, the greater the loss if a battle satellite should fail or if the Soviet Union should attack it successfully. Moreover, if too many interceptors were clustered together on single satellites, large areas might be beyond the maximum range of any interceptor at particular points in time. This consideration alone should limit the maximum number of interceptors located on one satellite to about 175 when only ICBMs are considered, and to about 250 when the entire presumed Soviet force of long- and intermediate-range missiles is the basis for planning.

If this maximum number of interceptors was placed on a single satellite, however, the Soviets could cut a hole in the boost-phase defense layer large enough for their land-based missiles by destroying as few as 9–12 battle satellites. To guard against this possibility, we have assumed that no more than 50 interceptors would be based on each battle satellite. By basing the interceptors in groups of 50 we have also guaranteed that enough interceptors would be close enough to the offensive missiles to make possible the timely arrival of the second wave of interceptors.

The Gamma system also would include decoy satellites, which would not contain interceptors or participate in the battle, but which would reproduce the radar and all other signatures of the real battle satellites. We have assumed that 5 decoys would be deployed for each real battle satellite and that they would be effectively indistinguishable from them. They thus would complicate proportionately any Soviet effort to destroy the defense. Readers should note that creating these decoy satellites might be particularly challenging, since the Soviet Union could presumably inspect all satellites from close range for long periods of time prior to an attack.

Given the assumed operational requirements, and assuming further that 10 percent additional battle satellites would be sufficient to compensate for missiles that are off-line due to system failures (awaiting maintenance), a total of 1,335 battle satellites and 6,675 decoys would be required to defend against Soviet ICBMs and SLBMs; a defense against the presumed complete force of long- and intermediate-range missiles would require 1,915 battle satellites and 9,575 decoys.

The battle satellites would be hardened to withstand attacks by ground-based lasers of current effectiveness.[26] We have assumed that this hardening, the use of decoys, and other measures would be adequate to protect the satellites from being attacked effectively. Here, again, we may have made an assumption that is optimistic for the defense.

Taking into account the functions required of the battle satellites to support the interceptors, as previously described, the battle satellites' need for capabilities to stabilize themselves and keep station, their need for a power source, and the systems redundancy necessary to give them high reliability for extended periods without maintenance, we have assumed that the total empty weight of the battle satellites would be equal to one-third their payload weight of 7,500 kg. This is about twice the weight of typical communications satellites used today; battle satellites, of course, would be more complex. This assumption also seems likely to be optimistic for the defense.

Battle Management Satellites

The battle management satellites would carry sensors and computers capable of detecting and tracking Soviet missiles and of projecting their trajectories with sufficient accuracy to allow the interceptors to be guided efficiently toward their targets. The sensors would include a radar with effective antijamming features, allowing instantaneous detection of missile launches and thus minimizing the defense reaction time.

These satellites would be netted together through high-capacity, antijam communications systems, so that operations requiring the involvement of more than one battle management platform could be coordinated effectively. Informal discussions with manufacturers suggest that with power generation and other "housekeeping equipment" included, these satellites might weigh approximately 20,000 kg.—roughly the size of the Skylab satellite.

The battle management satellites would be placed in circular orbits at an altitude of approximately 5,000 km. At this altitude a total of 16 satellites would be ample to maintain continuous surveillance of all the presumed launch areas. This number would allow some redundancy for an occasional off-line satellite.

The battle management satellites also would be hardened against attacks by ground-based lasers. Decoy battle management satellites, however, would not be deployed for two reasons. First, the number of battle management satellites would seem to be too low, and the likely cost of effective decoys, too high, for a survival strategy employing decoys to be cost-effective. Second, no anti-satellite systems capable of destroying satellites at the altitudes at which these satellites would operate have yet appeared, and it might yet be possible to agree on arms-control arrangements with the Soviets that would effectively block the development of such capabilities, though the prospects do not seem promising. If the development of such antisatellite systems was not effectively pro-hibited in future arms-control agreements, our assumption that the battle satellites would be survivable would likely be optimistic for the defense.

Launch Capabilities

The final integral component of the Gamma defense system would be the space launch capabilities necessary to place and main-tain the system in orbit. Assuming that battle satellite decoys would weigh 10 percent as much as their real counterparts (not including the on-board missiles), the total weight that would have to be placed in orbit initially to create a system to defend against the Soviet ICBM and SLBM force would be about 15.3 million kg. A system to defend against the entire Soviet long- and intermediate-range missile force would have an initial launch weight of about 21.9 million kg. Given the roughly 16,000 kg. payload the current space shuttle could deliver into polar orbit, launching these weights into orbit would require approximately 960 and 1,370 shuttle-equivalent flights for the two types of defenses, respectively.

We further have assumed that 10 percent of the initial launch weight would have to be orbited each year to allow repairs and a "rolling renewal" of the system. These repairs would be largely of the "remove and replace" variety and are thus assumed to require periodic visits by manned orbiters. If the current shuttle was employed and a full load of replacement materials was carried and used on every trip, approximately 100 and 140 shuttle flights per year would be required to sustain the smaller and larger of U.S. space-based defense systems, respectively.

On the basis of a recent study by R.G. Finke and others, the initial lift requirement for this system would not be great enough to justify the high development and associated infrastructure costs of the large unmanned booster one might otherwise choose for this task, so long as only the marginal costs of necessary shuttle flights would be incurred. Because a manned orbiter would be required for the sustaining lift tasks and could be used for the initial deployment, it would pay to upgrade the current shuttle with new pressure-fed liquid boosters. Since the *Challenger* accident makes some sort of upgrade necessary anyway, we have assumed that an improved shuttle would be available at no incremental cost to the Gamma system.[27]

Also according to the study by Finke et al., the upgraded shuttle would be capable of orbiting twice as much payload as the current shuttle. It could thus place about 60,000 kg. in an easterly low-earth orbit at an inclination of 28.5 degrees. The upgraded shuttle would be able to deliver about 32,000 kg. into the polar low-earth orbits required by the battle satellites.

Deploying this new launch system would require approximately 580 or 825 flights over a four-year period, plus approximately 50 or 70 flights per year thereafter, for the smaller and larger defense systems, respectively. If we take into account the reduced maintenance requirements during the first four years when the system would not have been fully deployed, an initial complement of 18 or 26 upgraded shuttles would be required, with the fleet tapering down to the 6 or 9 shuttles required to maintain the smaller and larger versions of the defense system, respectively. These figures are based on the Finke report, which assumes a maximum operations rate of 8 flights per year per shuttle and an effective lifetime for the improved shuttle of 100 flights.

Summary

Table 7 summarizes the incremental elements of the Gamma defense system, including launch requirements. The reader should recall that these elements would be added to the components previously deployed to create the Alpha and Beta systems.

TABLE 7
Incremental Elements of Strategic Defense System Gamma

Element	Number Required to Defend Against:	
	ICBMs and SLBMs	Long- and Intermediate-Range Missiles
Interceptors*	66,900	95,600
Battle Satellites*	1,335	1,915
Decoy Battle Satellites*	6,675	9,575
Battle Management Satellites	16	16
Upgraded Shuttle Flights (deployment)	580	825
Upgraded Shuttle Flights (maintenance) per Year	50	70

*A 10 percent margin is included for space systems awaiting maintenance (see page 70).

COSTS OF THE INCREMENTAL ELEMENTS OF THE GAMMA SYSTEM

We are now in a position to estimate the cost of each integral, incremental element of the Gamma defense system, as described in the following paragraphs.

Interceptor Missiles

The U.S. antisatellite interceptor missile (ASAT), now being developed as part of the "Space Defense System" by LTV, Inc., is a good technical analogue to the Gamma system's interceptor missile. The Gamma system's space-based interceptor would have a somewhat easier target to detect and track, which could make it cheaper to build than the ASAT, but the requirement that it remain reliable for long periods without maintenance could offset any such savings. The currently planned ASAT interceptors are expected to cost about $12 million each in the small quantities now being purchased. We have assumed that the first unit cost of the space-based interceptor would be the same—which seems likely to be a conservative assumption.[28]

Assuming, as we have done for missile costs estimated previously, that the marginal cost of producing the interceptors would decline by 10 percent for each doubling of the number manufactured and that 10 percent extra missiles would be sufficient for testing and a ground-based maintenance pool, the cost of producing the 73,600 interceptors necessary to defend against Soviet ICBMs and SLBMs would be about $180 billion. The cost of sufficient missiles to protect against both long- and intermediate-range ballistic missiles would be roughly $246 billion. These figures imply an average cost of $2.5 million and $2.3 million per missile for the smaller and larger defense systems, respectively. This average cost is about twice that of the air-to-air missile described as part of the Alpha system. That missile, however, would not carry nearly so capable and complex a "front end."

Consistent with our previous assumptions regarding the costs of developing new interceptor missiles, we also have assumed a research and development cost of $3 billion. While the space-based interceptor would be more complex than many other interceptors, its development program could take advantage of the experience gained with the current antisatellite weapon.

Finally, we have assumed that virtually no maintenance of these missiles would be done in space. Failed missiles would be replaced when the shuttle maintenance flight made its periodic visits to each battle satellite; the missiles would then be repaired on earth. Missile operating costs thus would consist primarily of spare parts and transportation to and from the host battle satellite. We have assumed conservatively that ten years of spare parts could be purchased for a price equal to one-half the original cost of the missile. The transportation costs will be accounted for in our overall estimate of space launch costs below.

Battle Satellites and their Decoys

Based on historical experience, the cost of developing the battle satellites would presumably be $2 billion. Informal discussions with manufacturers suggest that this kind of satellite could be purchased for roughly $35,000 per kg. We previously assumed that the battle satellites would weigh 2,500 kg., thus suggesting a total cost of about $85 million per satellite. This price would be somewhat less than current communications satellites cost typically.

Satellites are bought today in small numbers, they are built essentially by hand, and changes are often made from one satellite in a series to the next. Accordingly, the cost of today's satellites does not usually decline as manufacturing experience grows. In the event that large numbers of the same kind of satellite were produced, as would be required if the Gamma system was deployed, and if the design was "locked in" relatively quickly, the fairly long production run might allow a reduction of 10 percent in production costs for each doubling of the number of satellites produced.

Using $85 million as the first unit cost, the cost of producing enough battle satellites to defend against Soviet ICBMs and SLBMs would be about $42 billion. Similarly, the cost to purchase sufficient battle satellites to defend against both long- and intermediate-range missiles would be about $50 billion. These estimates imply an average cost per kg. of $10,500 and $12,500, respectively, for the battle satellites required by the smaller and larger versions of the space-based defense system. These "cost per kg." figures would be very low compared with today's satellite cost standards.

We have assumed that the satellites would require only the occasional replacement of major components, which would themselves be repaired on earth. We have assumed further that spare parts sufficient for ten years of operations would be purchased for a cost equal to one-half the original cost of the satellites. The cost of transporting replacement components to and from space is accounted for below.

We have assumed that effective decoy satellites could be developed, built, and maintained for 5 percent of the cost of real battle satellites. Since 5 decoys would be deployed for every real battle satellite, the total cost of purchasing and maintaining the decoys would be 25 percent of the cost of the real satellites.

Battle Management Satellites

The cost of research and development for the battle management satellites would be, in our assumption, $2 billion. On the basis of informal discussions with manufacturers, the cost of producing these relatively complex satellites should be roughly $50,000 per kg. We previously estimated that such satellites might weigh in the neighborhood of 20,000 kg., leading to an estimated cost of $1 billion each. This would be roughly 25 percent more expensive than press reports of the cost of current satellites incorporating advanced technologies that are used for military purposes. The battle management satellites, of course, would need to be exceedingly sophisticated in design and function. They would carry multiple sensors including a fairly sophisticated radar for instantaneous detection of Soviet missile launches.

To keep 16 satellites on line, a total of 20 would be bought for $20 billion. The number purchased seems too small for significant savings to result from the experience gained in building them. We have assumed that these satellites would be transported back to earth for repairs and that ten-year maintenance costs would be equal to one-half the original procurement cost of the satellite plus the transportation costs accounted for below.

Launch Costs

The marginal cost for launches of the upgraded shuttle previously described is calculated to be $57 million per launch, based on figures given in Finke et al. On this basis, it would cost about $33 billion to launch and sustain the smaller version of the Gamma defense system over the first four years of its deployment, and $47 billion for the larger version. Launch costs to sustain the two versions thereafter would come to $2.8 billion per year and $4.0 billion per year, respectively.

Summary

While the major costs of the Gamma defense system have been captured above, it is likely that a number of added, significant expenses not directly accounted for in the major components of the system would arise. For example, there would likely be additional costs to: (1) expand NORAD even further to manage this more complicated defense system, and (2) develop the required battle management software, which would be technically demanding and likely to absorb a significant portion of pertinent industries for years. Such costs are difficult to assess. We have included a nominal $5 billion procurement cost and annual $500 million operating cost to cover them.

Table 8 summarizes the costs of the incremental elements of the Gamma defense system. The total ten-year acquisition and operating cost for the space-based portion of the Gamma defense system would be about $460 billion if the system was intended to defend against ICBMs and SLBMs, and $600 billion for a system to defend against both long- and intermediate-range ballistic missiles. Roughly 60 percent of this cost would be attributable to interceptor missiles.

The complete cost of the Gamma defense system would include the cost of the underlying Beta system. The Beta system as previously described would be inappropriately configured for backing up the space-based defenses, however. The reason is that even without terminal defenses it would require more weapons to attack most of the strategic forces bases than would remain at the bases, simply because the space-based layer would charge a price of 100 attackers for every 1 that leaked through.

It thus would seem appropriate in the presence of a space-based defense component as effective as the one postulated for the Gamma system, to reconfigure the terminal defenses of the Beta system to provide less protection for strategic forces and a more uniform and heavier area defense to protect the population and other national values. Enough missiles are included in the Beta system to place a missile base containing 10 high endo-atmospheric interceptors within effective operating range of every point in the United States and most of southern Canada. The low endo-atmospheric missiles would be left in their original locations to

TABLE 8
Costs of Strategic Defense System Gamma
(billions of 1987 dollars)

*(Defense against ICBMs and SLBMs only/Long-
and intermediate-range ballistic missiles)*

Element	R&D	Procurement	Construction	Ten-Year Operations	Total Cost
Interceptors	3	180/246	—	90/123	273/372
Battle Satellites	2	42/50	—	21/25	65/77
Decoys	1	11/13	—	5/6	17/20
Battle Management Satellites	2	20	—	10	32
Upgraded Shuttle Flights (deployment)	—	33/47	—	—	33/47
Upgraded Shuttle Flights (maintenance)	—	—	—	28/40	28/40
Other	—	5	—	5	10
Subtotal	8	291/381	—	159/209	458/598
Beta (reconfigured)	15	81	11	62	169
Total	23	372/462	11	221/271	627/767
Rounded Total	**20**	**370/460**	**10**	**220/270**	**630/770**

provide a margin of greater protection for the most valuable of the strategic forces bases.

This rebasing operation should cost on the order of $3 billion, which is shown in Table 8 as an additional construction cost for the Beta addition to the costs of the space-based component of Gamma. Adding in the costs of the reconfigured Beta system, we obtain a total cost for the Gamma system of $630 billion or $770 billion, depending on whether the defense was intended to cover ICBMs and SLBMs alone or in addition to intermediate-range missiles.

4.
NOTIONAL STRATEGIC_____
DEFENSE SYSTEM DELTA
A Comprehensive Defense Using
Space-Based Laser Weapons

The cost estimates made for the Gamma defense system suggest that a space-based system employing interceptor missiles to destroy Soviet missiles during the boost phase or postboost phase of their trajectories would be very expensive. This is one reason why the current SDI research program is emphasizing directed energy weapons, at least for the space-based layers of any defense architecture.

ASSUMED CHARACTERISTICS OF A LASER DEFENSE SYSTEM

The advantage of directed energy weapons is that the range at which they can be effective is not limited in the same way that it is for missile interceptors, as travel time for the energy needed to destroy the target is nearly instantaneous. As was evident in our calculations for the Gamma system, the relatively lengthy travel times for the interceptors forced us to use a large number of battle satellites and a very large number of high-velocity missiles to ensure that an adequate number of interceptors would be within range of every Soviet missile no matter when it might be launched. These large numbers of interceptors and battle satellites drove the costs of the Gamma defense to very high levels.

The laser weapons discussed in connection with the Delta defense system are also limited in range, but much less so, and by a very different phenomenon. The practical limit on a laser weapon of the type that might be employed in a space-based defense would be determined by how tightly it could focus the light energy that it projected onto its targets. The minimum size of the spot of energy that could be generated would be determined by the diffraction limit of the optical system used to focus the light beam.

This diffraction limit determines a minimum angular spread to the beam that is proportional to the wavelength of the light and inversely proportional to the diameter of the focusing mirror. Because the light beam must spread out at, or above, this minimum angular spread, the minimum spot diameter that could be achieved would increase linearly with range. The limit on a laser's lethal range thus would be the maximum range at which the power in the laser beam could be focused on a target in a sufficiently small spot so that the total energy delivered within the spot during the period that the target could be illuminated is enough to damage the missile fatally.

Given this minimal description of how a laser defense would function, the primary objectives in designing such a defense can be readily understood. The laser should generate the maximum power possible; create light of the shortest possible wavelength; employ the largest possible mirrors; and have pointing and tracking systems that could allow the laser to shift its beam from target to target rapidly, focus quickly, and hold the focused spot on as small a total area of the missile as possible by minimizing jittterrrs [sic].

The means by which a missile could be protected against attack by laser weapons are similarly clear. A missile should have the minimum practical booster burn-time so that it would be exposed to the laser during this most vulnerable period of its trajectory as little as possible. Minimizing booster burn-time also would maximize the fraction of the boost phase that would take place at lower altitudes, where the atmosphere affords some degree of protection from directed energy weapons. The missile also could be made shiny to reflect away as much energy as possible, could be rotated continuously during launch to spread the attacking laser's energy over a larger area, and could be coated with a protective material.

None of these counteractions would be accomplished for free, although rotation of the missile during launch would be relatively inexpensive. None seems likely to afford complete protection from attack by the kinds of laser weapons the United States conceivably might build. Such measures, however, would require higher total amounts of energy to be delivered against each missile by the attacking laser and, all other things being equal, would increase the time that any given laser would have to illuminate a missile to destroy it. This, of course, would raise the minimum number of lasers required to deal with any given number of Soviet missiles.

Numerous estimates have been made in the unclassified literature for the values that conceivably might be achieved for the various parameters that define the potential effectiveness of a laser defense system and the potential vulnerabilities of future Soviet missiles. We assume the parameter values shown in Table 9. The values assumed for the laser system are at, or very near, the optimistic end of the spectrum of values described in publicly available sources.

TABLE 9
Assumed Characteristics of a Laser Defense System
and Its Target Missiles

Laser System:

Infrared laser light of 2.7 micron wavelength

Continuous power output of 25 megawatts

Sufficient fuel aboard to attack all assigned targets

Flawless focusing mirror of 10 meters diameter

.1 seconds required to slew, settle, and refocus on a new target

Negligible spot jitter

Target Missiles:

Hardness requiring delivery of 15 kJ/sq.cm. total energy

Booster effective attack exposure time of 80 seconds

The key assumption about Soviet missiles is that they would be hardened to withstand 15 kJ. (kilojoules) of delivered energy per sq.cm. This degree of protection from directed energy weapons lies about halfway between the degree of hardness assumed in the previously mentioned OTA report and the hardness that such outside experts as Richard Garwin assume to be feasible.[29] The degree of hardness would come at the expense of a significant fraction of the throw-weight of Soviet missiles. Although some of this throw-weight loss might be compensated for by improved warheads designed to allow a given yield to be achieved in a lighter package, the costs of hardening would still be significant.

Recognizing that atmospheric effects would reduce the effectiveness of the lasers against boosters at very low altitudes, we have reduced the effective attack exposure time of the missile by 20 seconds. Thus, 80 seconds is taken as the maximum period available for attacking the booster or its postboost vehicle, as compared to the 100 seconds utilized in the Gamma system.

With these assumptions about both the laser system and its targets, the laser would have to illuminate the missile effectively for a period of nearly 3.5 seconds to destroy a missile at a range of 3,000 km. If we ignore atmospheric effects, a single laser could destroy a maximum of approximately 25 missiles at this range.

At an altitude of approximately 800 km. a laser passing directly over an area within which missiles were boosting toward their targets would have many missiles to attack within much lower ranges, allowing smaller spot sizes and correspondingly higher energy densities to be achieved. It thus would require less time to destroy these closer missiles, and a higher potential number of successful attacks could be achieved by such nearby lasers.

COSTS OF THE DELTA SYSTEM

The costs of a space-based laser system would be determined mainly by the number of lasers required, the total mass of satellites that must be launched into space, and the cost of the individual laser satellites.

Garwin has estimated the number of required orbiting satellites for boost-phase intercepts for a wide variety of parameter values.[30]

By extrapolating from his figures, we have estimated that approximately 235 of the laser satellites described above would be necessary to defend against the projected force of Soviet ICBMs and SLBMs and that about 335 such satellites would be required to defend against the entire force of long- and intermediate-range missiles previously described.

These satellites would be placed in orbits inclined at 60 degrees to the equator. This would be optimal for defending against land-based missiles in the deployment area we have described. Even so, in these orbits each satellite would spend about 15 percent of its time with a line-of-sight range to the North Pole that would be within its maximum effective range. This implies that the laser satellite constellations also should be capable of dealing with SLBMs launched from the area of the North Pole. For the same reasons argued in the Gamma case, we have assumed that the Soviets would not attempt to overwhelm the defense system with potentially risky tactics, such as moving large numbers of strategic submarines to a small area near the equator for a concentrated, simultaneous launch.

Readers should note also that in the Gamma system, polar orbits had to be employed because of the short ranges of the missile interceptors. The longer effective ranges of the laser satellites would allow them to be located at considerably lower inclinations and, accordingly, to concentrate their defensive capabilities against the more numerous and, by assumption, more densely deployed land-based missiles. Moreover, with very capable lasers (of the kind we are considering here) shooting at long average ranges, the increase in the number of lasers required to cope with reductions in the offensive missile deployment area beyond the factor of three assumed in this analysis would be very slight.

To determine how much mass must be put in space to deploy the Delta system, we must now estimate the weight of the individual satellites. Garwin and Bethe double the weight of fuel required for a 25-megawatt laser to operate for 100 seconds, to create a "bare minimum" weight estimate of 10 tons per satellite.[31] Industry estimates of total weights for laser satellites run even much higher. *Aviation Week & Space Technology* gave a 100-ton estimated weight for a laser satellite with similar power generation capabilities.[32] Reflecting these various estimates, we have assumed optimistically a total of 40 tons for the weight of laser satellites in the Delta system.

We have assumed that some means would be found to harden adequately or otherwise defend these satellites against Soviet attacks. One conceivable means might be to enclose the satellite in a shell of material that could absorb directed energy from an enemy's laser for some short period of time. This period would be employed to move additional protective material into any areas on the shell that had come under attack. We have assumed that the weight of suitable, movable "armor" and/or other defenses would be included in the 40-ton total weight assumed for the laser satellites.

Given a weight of 40 tons per satellite, the total weight of the laser constellation required to defend against the projected Soviet ICBM and SLBM force alone would be 8.5 million kg. The weight of the constellation needed to defend against the entire projected force of Soviet long- and intermediate-range missiles would be 12.2 million kg.

On the basis of these assumptions, the costs for those elements of the Delta system that would be different from the corresponding elements of the Gamma defense are estimated below. All other elements of the Gamma system, including the battle management satellites, have been assumed to be the same.

Laser Satellites

As noted previously, the costs of satellites with generally similar functions tend to scale with their total weight. Thus, if we assumed that the costs of the power and laser light generation systems were roughly the same per kg. as those for the required optics, we could project a rough first unit cost for the laser battle satellites from the costs of the Hubble Space Telescope. This latter satellite weighs approximately 12 tons and costs $1.2 billion. Scaling this cost linearly would suggest a first unit cost of $4 billion for Delta laser satellites.

Alternative means of estimating the laser satellite's cost are available, however. For example, if we applied a cost factor of $60,000/kg. (the average of the per kg. costs for a typical communications satellite and the Apollo vehicle), we would obtain a total cost of $2.2 billion for the 40-ton laser satellite. Finally, informal discussions with manufacturers suggest a first unit cost of $1.75 billion.

We have taken this last figure as the first unit cost for the laser battle satellite. Assuming that each successive doubling of the number of satellites produced would lead to a reduction of 10 percent in their marginal cost, the total costs of manufacturing the 235 satellites required to defend against Soviet ICBMs and SLBMs would come to about $210 billion. The constellation for defending against the projected force of all Soviet long- and intermediate-range missiles would cost $280 billion. In recognition of its enormous complexity, we also have assumed a development cost for this laser system of $15 billion. If we further assumed that an average of only 5 percent of the components of these satellites would need to be replaced each year, spare parts requirements would come to roughly another $10.5 billion per year for the smaller constellation, and $14 billion per year for the larger one.

Initial and Sustaining Launch Costs

The total weight that must be launched into space initially for each of these systems, including the 16 battle management satellites, would be approximately 8.8 million kg. for the smaller system and 12.5 million kg. for the larger one. If we assumed that an average of 5 percent of the satellites' weight would have to be launched each year to maintain the system, another 440,000 kg. per year would have to be launched to support the smaller system, and approximately 600,000 kg. per year for the larger. As with the Gamma system, we have assumed that upgraded shuttles would be employed to activate and support the laser constellation.

The upgraded shuttle is estimated to be capable of delivering 60,000 kg. into an eastward low-earth orbit at our desired inclination of 60 degrees. Roughly 145 flights would thus be required to place the smaller laser constellation in orbit; roughly 210 would be required for the larger constellation. Similarly, assuming that the satellites would fail so infrequently as to make most of the maintenance preventative, roughly 7 flights per year would be required to maintain the smaller constellation, and 10 flights to maintain the larger.

If we again used a four-year deployment period, 5 upgraded shuttles would be adequate to deploy, activate, and maintain the smaller constellation during this initial period; about 7 would be required for the larger. Again employing the estimates in Finke et al., the launch costs to deploy the smaller constellation would be approximately $9.4 billion and the sustaining launch costs would be roughly $415 million per year. The corresponding launch costs for the larger constellation would be $13.5 billion and $590 million per year.

Total Costs

By replacing the battle satellites and associated launch costs of the Gamma system with those just computed for a space-based laser system, we obtain the costs of the Delta system summarized in Table 10. Comparisons of these figures with the costs estimated for the Gamma system in Table 8 indicate that a space-based laser system of the Delta kind would be roughly 80–85 percent as expensive as a space-based missile system of a similar overall effectiveness.

We should not conclude, however, that a directed energy system would not ultimately prove to be even less costly. The technology of directed energy systems is far less mature than that for interceptor missile systems, and our sketch of how well a laser system might work, what its components might have to weigh, how many might be needed, and how much each one might cost are accordingly much less certain than for a missile interceptor system. The message that may be drawn from these estimates is that in designing a research strategy for strategic defenses, it may be preferable to aim for a directed energy system with considerably better performance parameters than we have assumed.

TABLE 10
Costs of Strategic Defense System Delta
(billions of 1987 dollars)

*(Defense against ICBMs and SLBMs only/Long-
and intermediate-range ballistic missiles)*

Component	R&D	Procurement	Construction	Ten-Year Operations	Total Cost
Laser Satellites	15	210/280	—	105/140	330/435
Battle Management Satellites	2	20	—	10	32
Launch Costs	—	9/13	—	4/6	13/19
Other	—	5	—	5	10
Subtotal	17	244/318	—	124/161	385/496
Beta (reconfigured)	15	81	11	62	169
Total	32	325/399	11	186/223	554/665
Rounded Total	**30**	**330/400**	**10**	**190/220**	**550/670**

5.
COSTS OF EUROPEAN COMPLEMENTS FOR A COMPREHENSIVE STRATEGIC DEFENSE SYSTEM

In THE EVENT THAT THE UNITED STATES decided to deploy a comprehensive strategic defense system, it must be assumed that the Soviet Union also would make a maximum effort to deploy an effective system with a comparable objective. Although the Soviet defense system may not appear on exactly the same schedule or work in exactly the same way as the U.S. system, the prudent assumption would be that comparably effective Soviet defenses would exist not very long after the U.S. system had been deployed fully. Thus, a situation would have been created in which both sides' offensive nuclear forces would have only very limited capabilities against one another.

It also seems likely that even if the United States was itself secure from nuclear attack, it would wish to continue to maintain its security commitments to the nations of Western Europe, to Japan and Korea, and to other states—and would seek to maintain military capabilities sufficient to deter an adversary from challenging those commitments. One consequence of the deployment of effective nuclear defenses by both the United States and the USSR would be removal of the threat of long-range nuclear strikes as a deterrent to the initiation of warfare in Europe and other regions; this threat, of course, is a key element in NATO's political/military strategy.

Three major changes, accordingly, would follow logically for NATO's force posture in the event of deployment of the Gamma or

Delta strategic defense systems. First, ground-based missile interceptor systems and additional air defenses would be deployed in Europe to back up the space-based component of the system in defending the continent against attacks by long-range, nuclear-armed missiles and aircraft. These defensive systems, together with the space-based component, could reduce sharply the probability of penetrations in depths of hundreds of kilometers or more by aircraft or missiles and, accordingly, raise the perceived value of shorter-range nuclear forces. Second, NATO could strengthen its own shorter-range nuclear forces sufficiently to establish and maintain a balance of such forces with the Warsaw Pact and thus to deter their use. Third, to compensate for the effective loss of the deterrent value of long-range nuclear forces, NATO could improve its capabilities for conventional warfare substantially.

These assumed improvements in the NATO force posture follow logically from our assumptions about long-range strategic capabilities. They might also be necessary in the absence of U.S. deployment of a comprehensive strategic nuclear defense system (if the Soviet Union deployed such a system and the United States did not, for example, or if the Soviet Union gained a significant edge in offensive nuclear capabilities), but they would certainly be necessary in the circumstances postulated above. How far each improvement might be taken is difficult to forecast.

Moreover, in some cases European defenses could not take the same form as those for the United States because of underlying geographical differences. The flight times to Western Europe of missiles launched from Warsaw Pact territory would be much shorter than those of long-range missiles flying from the Soviet Union to the United States, for example. Similarly, air defenses for Europe could not have depth comparable to that available for the defense of North America.

For these reasons, we have described only nominal improvements in West European military capabilities that would follow logically from the deployment of effective, comprehensive air and missile defenses for North America, and we have estimated their costs very roughly. These estimates will give readers a benchmark for understanding the full range of costs that might be entailed in defending North America comprehensively against nuclear attacks.

The costs of representative NATO defense improvements, as well as the extra expense of giving the space-based layer the additional capabilities necessary to handle longer-range missiles aimed at Europe, are listed separately in our tables so that readers can adjust their magnitudes or decide not to consider them at all in the cost of strategic defenses.

We have not made any estimates of the cost of additional military capabilities that might be required to defend Israel, Japan, Korea, or other U.S. allies if both the United States and the Soviet Union deployed effective air and missile defenses.

AIR AND TERMINAL MISSILE DEFENSES IN EUROPE

The deployment of ground-based air defenses and terminal missile defenses in Europe would provide: (1) a final line of defense against intermediate-and long-range ballistic missiles, (2) protection against short-range ballistic missiles, and (3) defenses against tactical and long-range aircraft. Each of these missions would pose distinctly different problems.

Given the existence of space-based missile defenses, gaining added protection from intermediate- and long-range ballistic missiles would be relatively straightforward. If, for example, as many as one-half of the entire Soviet force of ICBMs, SLBMs, and IRBMs were launched against Europe and a space-based defense as capable as the one postulated in the Gamma system was in place, only 15 missiles would be expected to survive through the boost phase and challenge the ground-based segment of the defense. If the surviving missiles were distributed proportionally by type, as they were in the original force, the terminal defenses would have to contend with approximately 100 warheads.

If the light area defenses included in the Beta system also had been deployed in Europe, they could destroy 90 percent of the leakers that attacked defended areas. If a two-layer defense of the kind incorporated in the Alpha system had been deployed in Europe, it could ensure that no more than 1 warhead could be expected to threaten a relatively hard target, such as an airfield or a protected command and control center.

With these proportions in mind, we have assumed that as a complement to the Gamma or Delta systems, a light area defense would be provided for all thirty-three of the urban areas with populations in excess of 1 million that are located on the territory of the European members of NATO. We have assumed that the average area that would have to be protected would be 350 sq.n.mi., the same assumption made for the U.S. deployments of light area defenses. To calculate the number of ground-based interceptor missiles required to protect these urban areas, as well as the number of individual locations (bases) at which the missiles would be deployed, we have also followed the method used in the U.S. case, with the result that 1,160 missiles at 97 bases would be required.

In the European case, however, it also would be necessary to purchase laser radar/battle management aircraft of the kind included in the Alpha system to control the intercepts. Six such aircraft would be sufficient to cover the area to be defended. We have assumed that double coverage would not be planned, as in the U.S. case, but sufficient aircraft would be purchased to allow a continuous airborne alert in times of crisis, thus reducing the potential for a disastrous preemptive attack by the Soviet Union. Assuming that only 20 percent of the aircraft could be maintained on airborne alert, a total force of 30 laser radar/battle management aircraft would be required.

Employing the same cost estimation techniques used for the Alpha and Beta systems, it would cost $2.8 billion to purchase the required ground-based interceptor missiles and construct their bases and about $140 million to operate the missiles each year. It would cost an additional $7.5 billion to purchase the required radar/battle management aircraft and an additional $450 million per year to operate them.

To provide additional protection for particularly important military targets and to blunt short-range missile attacks against such targets close to the border between NATO and the Warsaw Pact (or close to ocean areas from which the Soviets might launch short-range missiles from submarines), we also postulate that NATO would construct two-layer ground-based defenses of 30 key hardened targets. The low altitude terminal defense would be similar to the lower layer interceptors employed in the Alpha system.

These defenses would employ a total of 3,000 HEDI missiles, 300 LEDI missiles, and 30 radars, at 50 missile bases. Again utilizing the same estimating techniques employed in case of U.S. defenses, the total acquisition cost of the missiles, radars, and basing installations required for this two-layer defense system would be $5.4 billion; it would cost about $340 million to operate this component each year.

Two additional components would be deployed to stiffen NATO's air defenses. Because Western Europe does not have sufficient airspace forward of its highest value targets to mount the kind of air defense in depth postulated for North America, we have assumed that additional surface-to-air missile systems would be deployed to create a new air defense belt in the eastern portions of the Federal Republic of Germany. Two hundred Patriot-type missiles and launchers would be acquired for this purpose, as would 20 mobile air defense radars. The procurement cost of these components would be approximately $750 million. Operating costs would be roughly $250 million each year.

The second air defense system would consist of 40 armed surveillance aircraft of the type employed in the Alpha system. These aircraft would be put on airborne alert during crises. While they could not comprise a uniform area defense, or even sustain a single continuous barrier, 40 such aircraft would be sufficient to maintain an airborne alert over an area about two-thirds the size of the European portion of NATO or to maintain continuously a barrier approximately 6,000 n.mi. long. This complement of aircraft together with their missiles would cost about $9 billion to purchase and $500 million to operate each year.

Finally, we have assumed that both the radar/battle management aircraft to control missile intercepts in the outer layer and the armed surveillance aircraft for air defenses would be deployed on NATO's current air bases. Additional hangars, maintenance facilities, and taxiways would have to be added, however, and would cost about $600 million to construct. Base operating costs also would be greater because of the larger number of aircraft to be deployed; the incremental expense would be perhaps $120 million per year.

The total cost, then, of the complementary improvements to NATO's air and missile defenses in Europe that would follow logically from the deployment of space-based missile defenses would equal about $26 billion in procurement and construction expenses, plus $1.8 billion per year in operating costs. Total ten-year systems cost would be roughly $44 billion.

IMPROVEMENTS TO SHORT-RANGE THEATER NUCLEAR FORCES

Since deployment of effective air and ballistic missile defenses would imply the neutralization of NATO and Warsaw Pact intermediate- and long-range nuclear forces, it would be essential for the Alliance to maintain a rough balance of military capabilities with the Warsaw Pact at each potential lower level of warfare. Otherwise, with the nullification of NATO's deterrent strategy based on the ultimate threat of nuclear strikes on Soviet territory, the Soviet Union might perceive opportunities to seize NATO territory or to disarm NATO's armed forces and that to do so implied little countervailing risks. In recent years, with its deployment of SS–21, SS–22, SS–23 short-range missiles, and with apparent preparations for the use of nuclear artillery and rockets in the event of conflict, the Warsaw Pact has closed what had been a long-standing NATO advantage in battlefield nuclear weapons, and it now threatens to gain a clear lead.

In the event that the United States deployed a comprehensive strategic defense system, it thus would be desirable for NATO to deploy additional tactical nuclear weapon systems to compensate for future Pact capabilities. There would be many alternative means of accomplishing this objective. Notionally, however, such a program might consist of the incremental deployment of 1,500 155-mm. artillery shells and 300 modernized "Lance" missile launchers, each with 5 missiles. Such deployments would greatly enhance NATO's tactical nuclear capabilities, thus strengthening the deterrence of any Soviet first use of nuclear weapons on the battlefield. It should be noted, however, that placing greater emphasis on short-range nuclear weapons for the defense of Western Europe would require a major redirection of trends in NATO policies, which might be difficult to implement for political reasons.

It would cost approximately $5.5 billion to acquire these weapons. Their impact on future operating costs would be small, assuming that this modernized force could be maintained and operated by approximately the same number of people now required to operate the current force.

IMPROVEMENTS TO NATO's CONVENTIONAL FORCES

Most important, in the event that the Warsaw Pact deployed effective defenses against intermediate- and long-range nuclear forces, the deterrence of war in Europe would depend even more heavily than today on the Alliance's conventional military capabilities. How NATO might go about strengthening its conventional capabilities—what specific forces might be expanded, what weapons might be improved, how the burden might be shared among the members of the Alliance—is difficult to predict. For the purposes of this estimate, however, we have considered three possible steps.

First, NATO could accelerate and expand the development and deployment of nonnuclear weapons employing advanced technologies. According to a recent study by a group of eminent retired military and civilian officials, and private experts, such technologies could be used to improve NATO's capabilities in six important ways: (a) to identify preparations for attacks and acquire targets deep in enemy territory, (b) to blunt the initial attack, (c) to counter the Pact's growing tactical air power, (d) to disrupt "follow-on forces," (e) to disrupt Pact communications, and (f) to improve NATO's own command, control, and communications capabilities. The panel estimated that a program to implement these improvements would cost, at the outside, approximately $30 billion at 1987 prices.[33] General William Rodgers, the supreme allied commander in Europe, has endorsed these proposals and indicated that in his view the cost estimate was "realistic."[34]

Second, the Alliance could improve the capabilities of its existing ground and air forces to fight a conventional war. Steps might include the construction of physical barriers across likely attack routes; improvements to various logistical facilities; and expansion of the stocks of war munitions, spare parts, and other equipment available to allied forces from a level necessary for fifteen days of warfare to one offering a thirty-days' supply. William Kaufmann has estimated that these three measures would cost about $40 billion (at 1987 prices).[35]

Third and finally, NATO might consider expanding the size of the ground and air forces it maintains in Europe. Any such step would be difficult for the Alliance, not only because of existing constraints on defense budgets, but also because each of the major Alliance members faces tightening constraints on available workers during the remaining years of the century. In the case of the United States and Great Britain, both of which have voluntary armed forces, the maintenance of existing military personnel levels is expected to become increasingly difficult. Any substantial increase in the size of active-duty forces would probably require either the reimposition of conscription or very sizable increases in military pay rates. On the other hand, the Federal Republic of Germany and France already rely on conscription and would have to extend their terms of service still further, and make other changes, to increase the size of their armed forces.

For these reasons, unless there was a massive deterioration in the state of East-West relations and a real threat of war, it is probably unrealistic to imagine an increase in size of the Alliance's active-duty forces. It may be possible to expand reserve forces, however, and maintain them in a sufficient state of readiness so that they could operate in the war zone in about thirty days. William Kaufmann has estimated, for example, that two additional German reserve divisions and one additional French reserve division might be established. It would probably cost about $22 billion to recruit, equip, train, and maintain these divisions for a ten-year period. Kaufmann also has suggested the possibility of establishing nine additional wings of U.S. close support aircraft; the ten-year cost of this step would be roughly $20 billion.[36]

It should be reemphasized that these steps are only illustrative of the types of measures that could be taken to improve NATO's capabilities to fight a conventional war. Many officials and experts believe that such steps should be taken under current circumstances. The case for them would be strengthened substantially if there was serious prospect that the Warsaw Pact would deploy effective defenses against NATO's nuclear deterrent.

TOTAL COSTS

Table 11 summarizes the costs of these complementary improvements in military capabilities in Western Europe, which would be necessary if effective missile defenses were deployed by NATO and the Warsaw Pact. All told, ten-year systems costs could reach $160 billion, in addition to the costs necessary to "thicken" either the Gamma or Delta space-based component sufficiently to defend against Soviet intermediate-range ballistic missiles.

TABLE 11
Potential Costs to Maintain an Effective NATO Military
Posture in the Event Comprehensive Defenses Are Deployed
(billions of 1987 dollars)

Type of Measure	Procurement and Construction	Ten-Year Operations	Total Cost
Terminal Missile and Air Defenses	26	18	44
Short-Range Nuclear Forces	6	negligible	6
Conventional Forces	82	30	112
Total	114	48	162
Rounded Total	**110**	**50**	**160**

NOTES: The costs of thickening the space-based component of any comprehensive strategic defense system to protect against Soviet intermediate-range ballistic missiles would of course also be relevant for the defense of NATO. These costs are presented as optional costs in the chapters on the Gamma and Delta systems.

6.
CRITICAL ASSUMPTIONS AND SENSITIVITIES

NUMEROUS ASSUMPTIONS WERE MADE in estimating the costs of the four notional strategic defense systems. Table 12 lists the more important of these assumptions. The optimistic assumptions are those that tend to reduce the cost estimates; pessimistic assumptions are those that tend to increase costs.

The optimistic assumptions primarily concern factors over which the United States may have the greater leverage. For example, while missile probabilities of kill depend in part upon how the Soviet Union designs its targets, the United States has an opportunity to prevail in this competition with very clever missile designs based on superior technology.

Even so, two of our optimistic assumptions pertain to factors that would be under Soviet control. First, we assumed that the Soviets would not increase their cruise missile threat because the air defenses we had postulated would make this competition a losing game for them. Given the quality of the postulated defenses, it would cost the Soviets more to increase their air threat than it would for the United States to offset the Soviet action with an expansion of U.S. air defenses.

The second and more important exception is our assumption that the Soviet Union's antisatellite effort could be countered effectively for costs on the order of those described for the satellites used in the systems. There is little basis for this assumption. The potential vulnerability of defense satellites to attacks appears to be

TABLE 12
Critical Assumptions

Alpha and Beta Systems

Optimistic Assumptions

Missile characteristics: cost, reliability, probability of kill
U.S. missile sensors would be effective despite Soviet efforts at concealment
Antisubmarine sensors would give reliable warning of Soviet submarines deployed close to U.S. shores
Battle management of missile and air defenses would be efficient
Soviet Union would not increase the air threat substantially
Production costs would continue to decline with experience

Pessimistic Assumptions

Soviet bombers would launch cruise missiles from outside the effective range of U.S. air defenses
Soviet decoys would make attacks outside the atmosphere infeasible

Gamma and Delta Systems

Optimistic Assumptions

Soviet antisatellite systems could be rendered ineffective at reasonable cost
Interceptor missile characteristics: cost, reliability, probability of kill, and weight
Laser characteristics: cost, power, retarget time, reliability, and weight
Only short decision delays would be involved before initiating attacks on missiles in the boost phase
Soviet submarines would not mass to shoot
Production costs would continue to decline with experience

Pessimistic Assumptions

Soviet boost times could be reduced by approximately 50 percent
Soviet land-based missile deployment area would be reduced by approximately two-thirds
Soviet missiles could be hardened to 15kJ/sq.cm. against laser attack
Decoys would make space-based defenses ineffective in midcourse

perhaps the largest unsolved problem for strategic defenses, and we are unaware of any promising solutions to it.

Another optimistic assumption that has particularly concerned some readers is our assumption that the various antiballistic missile interceptors employed in the several defense systems we have constructed would be capable of 90 percent single-shot probabilities of kill. A lower figure would clearly increase costs substantially if the overall performance of the defense systems was to be held constant.

If, for example, the HEDI and LEDI missiles of the Beta system had single-shot probabilities of kill of 80 percent instead of 90 percent, the cost of the Beta defense would rise by approximately $30 billion, or nearly 20 percent. If the single-shot probability of kill of the space-based interceptor was 80 percent rather than the 90 percent assumed in the estimates, a Gamma system of equivalent overall effectiveness would cost approximately $190 billion more for the smaller constellation and approximately $250 billion more for the larger. If all three of the interceptor missiles employed in the Gamma defense were degraded to 80 percent single-shot kill probabilities, maintaining constant overall performance of either version of this system would raise the costs by about 30 percent. This would bring the cost of the thicker system intended to defend against IRBMs, as well as ICBMs and SLBMs, to more than $1 trillion.

The pessimistic assumptions primarily concern factors over which the Soviet Union seems likely to have the greater leverage. For example, the Soviet Union alone would decide whether to move its missiles into a significantly smaller deployment area; the temptation to do so would be substantial. The costs of space-based defense systems such as those we have described would rise dramatically if the Soviets reduced their deployment areas sharply. Even if the Soviets showed no interest in doing so and agreed formally not to, the United States would nonetheless have to be prepared to cope with such a move.

Obviously, if the U.S. technical/industrial community was able to find relatively inexpensive solutions to the problems reflected in our list of optimistic assumptions, our notional defense systems could be built for less than the costs we have projected. Similarly, if our pessimistic assumptions proved too pessimistic, the cost of strategic defenses could be lower.

Regarding the pessimistic assumptions in particular, our assumption that Soviet bombers would launch their cruise missiles outside the range of U.S. air defenses seems to be widely shared in the defense analysis community. Our assumption that the use of lightweight decoys would make an effective defense impossible during the midcourse phase of the attacking missiles' flight is less widely shared. In fact, the current SDI program is devoting a great deal of effort to solving this problem.

Toward this end, SDI research generally is seeking active means of discriminating during midcourse between the reentry vehicles and lightweight decoys. One potential scheme was reported in *Time* magazine.[37] It would employ a space-based, neutral particle-beam generator to illuminate both decoys and targets. When particle beams strike a light object there is little or no interaction, but when a massive object is encountered, gamma rays are given off, which can be detected.

The task of building an effective ballistic missile defense could be eased enormously if an effective midcourse discrimination technique could be found. Instead of having on the order of 100 seconds in the boost phase, plus perhaps 30 seconds in the terminal phase of a ballistic missile's flight, to destroy the attacking warheads, as much as thirty minutes could be available. Extra time could allow more shots by individual defense satellites, and it would allow a much higher fraction of the satellites in a defense constellation to enter the battle.

The remaining pessimistic assumptions all relate to how the Soviet Union would design and deploy its land-based missiles. The particular assumptions we have made raise the cost of the Gamma system by a factor of about eight compared to the cost of using the same defensive systems to defend against today's Soviet missile force. Similarly, because the laser defense of the Delta system would be sensitive to hardening of its targets and to reductions in booster burn-times, our estimates of the Delta system's costs are at least five times greater than for a system adequate to defend against today's Soviet missile force.

We were not as pessimistic in our estimates as might seem reasonable. Soviet missile boost times could be made a factor of two shorter than we have assumed. The size of the deployment area could be decreased by another factor of ten or more, which would have a disproportionately large effect on the cost of the Gamma system but only a small effect on the cost of Delta. With perhaps another 10 percent sacrifice in payload weight, future Soviet missiles could be made another factor of two harder than we have assumed. These steps are completely under Soviet control and, in combination, could more than triple the overall costs of either the Gamma or Delta systems.

Readers also should appreciate the significance of our assumptions regarding the effects of "learning curves" on costs. Some analysts we have consulted argue that continuous design changes, among other factors, nearly always prevent the realization of significant cost savings as a result of manufacturing experience, at least for a system more complex than an aircraft cannon round. Others have argued that our assumption of a 90 percent learning curve is pessimistic; that the importance of sticking to one design would prevail in this case and that costs would accordingly drop more rapidly as large numbers of weapons were produced. During World War II, for example, the United States realized 80–85 percent learning curves in the long construction runs for combat aircraft. More recently, an 85 percent learning curve was experienced in manufacturing the external tanks of the shuttle.

We cannot be certain whether significant learning effects would be realized in building a strategic defense, of course, but if they were, the effects could be dramatic. The $370 billion estimate for interceptor missiles for the larger version of the Gamma defense system would be reduced to approximately $165 billion if costs could be reduced by 15 percent for each doubling of the number of missiles produced (rather than by 10 percent, as assumed in our estimates).

Finally, we note that the SDI program appears now to be placing great emphasis on exploring a newer strategic defense concept. This newer concept would use space-based mirrors to reflect and focus the energy of powerful ground-based lasers on their targets. This concept could reduce several of the problems illustrated in the analysis of the Delta system. First, by keeping them on the ground, much more powerful lasers could be employed. Second, by using mirrors, a relatively small number of lasers would be sufficient. Third, because more powerful lasers would be employed and because they would be on the ground, the satellites attacking the targets could be lighter and fewer in number. This would reduce the weight of material that would have to be placed in orbit.

This concept has its own unique problems, of course. It remains to be seen whether these problems, as well as those such a concept would suffer in common with the Delta system (such as the potential vulnerability of satellites to attack), will prove solvable.

PART 2

STRATEGIC DEFENSE
COSTS IN PERSPECTIVE

$\mathbf{T}$HE ESTIMATED COSTS OF THE FOUR notional strategic defense systems, including both total investment and ten years of operations, would range between $160 billion and $770 billion at fiscal 1987 prices. These aggregate figures indicate relatively little about the prospective fiscal and economic impact of a decision to deploy strategic defenses, however, as they do not address the key question of how costs might be spread over time. (The cost estimates used in this section exclude the cost of complementary measures that logically would follow from some of the alternatives, such as increases in NATO's military capabilities, but they include the cost of "thickening" the comprehensive defenses to protect against Soviet intermediate-range missiles. It seems logical to divide the effort between the United States and its allies in this fashion.)

This key question, in turn, would depend largely on the pace at which the strategic defense system would be deployed. In describing each notional defense system, we made rough assessments of when technologies might become available for deployment. Based on these assessments and on typical historical patterns of major weapon acquisition programs, figures 1 through 3 show how expenditures might be distributed over time for the Beta, Gamma, and Delta systems, respectively, following a decision in fiscal 1987 to develop the relevant system. Each figure also shows how the investment portion of each system's total cost (research, procurement, and construction) might be distributed. (The Alpha system would represent only a slight departure from the Beta cost distribution.) These cost schedules, of course, are only illustrative of how the notional systems might be deployed. Any of the systems could be acquired over a somewhat longer period of time to reduce their yearly cost—or, if warranted by events, schedules could be compressed somewhat, although this might mean an increase in real costs.

We have assumed that the Alpha and Beta systems (Figure 1) would be designed and developed between 1988 and 1995; construction of necessary air bases and other installations could begin in 1994, and procurement of necessary missiles, radars, and aircraft initiated in 1995. Given that these two systems would make use of technologies already in hand or close to having been developed, they could certainly be deployed on a more rapid schedule than this. That would probably be less efficient, however, and entail greater costs than those in our estimates.

On the assumed development and procurement schedule, deployment of either the Alpha or Beta system could probably begin in 2000; both could be fully operational by 2005. If the systems were deployed on such a schedule, budgetary demands for either system would build slowly over the fiscal 1988–97 period. During their peak ten years of funding requirements, fiscal 1997 through 2006, the two systems would require annual appropriations on the order of $10 billion and $11 billion, respectively. Costs would then drop to the annual operational expense of about $6 billion in either case. In rough terms, investment costs would account for two-thirds of the cost of either system, about $90 billion.

For the Gamma system (Figure 2) we have assumed that the components of the Beta system would be deployed on the schedule described previously and that the incremental space-based components would be developed between 1995 and 2002 and procured between 2000 and 2015. This relatively rapid schedule is possible insofar as the Gamma system also would make use of technologies that are close at hand, although developing a miniaturized "front-end" as capable as that envisioned for the space-based interceptor would be challenging. (The costs shown in the figures for both the Gamma and Delta systems, as well as those cited in the remainder of this section, pertain to the variant of the relevant system in which the space-based component had been thickened to permit

FIGURE 1: Annual Cost of Beta System

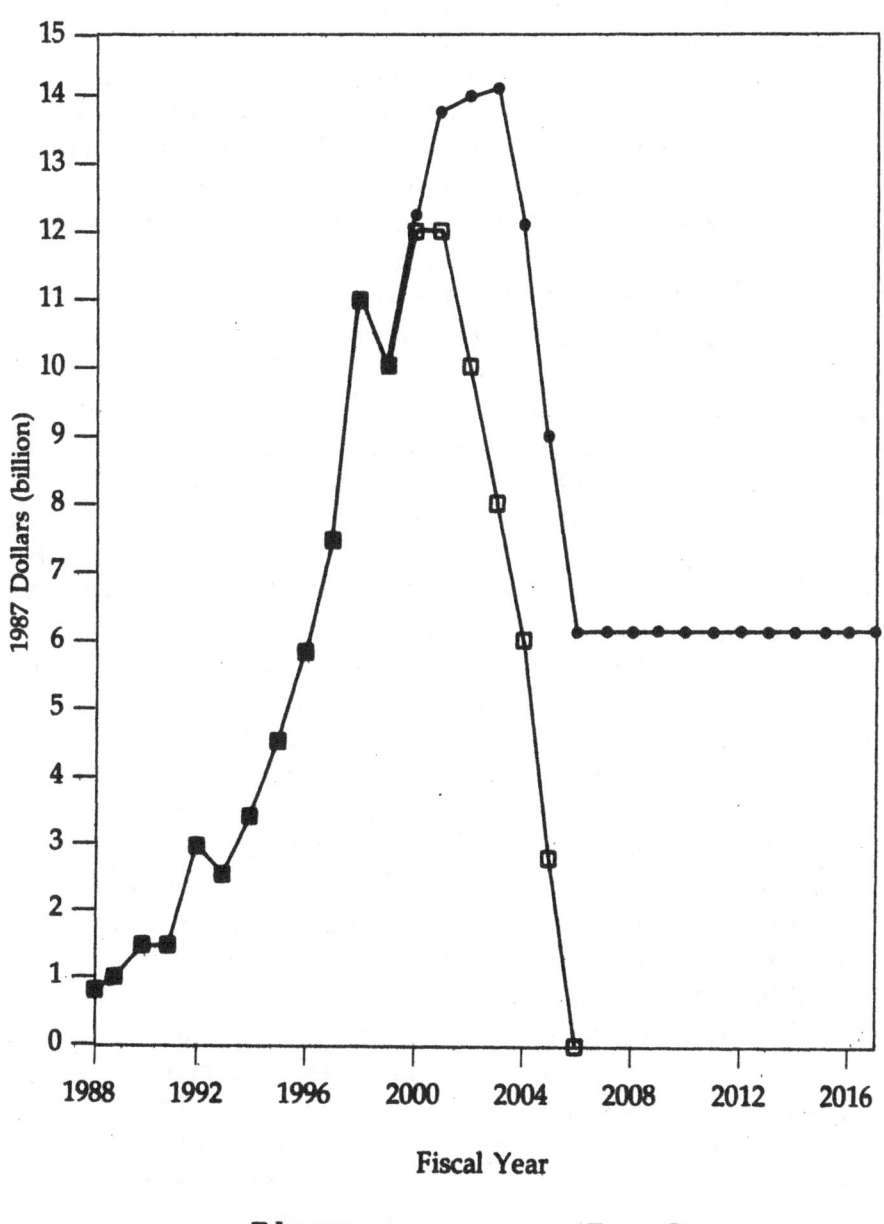

the defense to operate effectively against intermediate-range ballistic missiles, as well as against intercontinental land-based and submarine-launched ballistic missiles.)

We have assumed that deployment of the space-based components would begin in 2008. An initial operational capability then might be achieved in 2010, and the full system could be operational in the year 2012. On such a schedule the budgetary requirements for strategic defenses would mount rapidly after the year 2000, fluctuating between $35 billion and more than $50 billion annually during the period 2002 through 2014. During the peak ten years of funding for the Gamma system, 2005 through 2014, annual appropriations would average around $44 billion. After 2015 a continuing annual commitment of nearly $30 billion would be required to operate the system. Like the previous systems, investments for the Gamma system would account for nearly two-thirds of total costs.

Because it would assume the solution of more difficult technological problems, the Delta system (Figure 3) could probably not be deployed until the second decade of the new century. We have assumed in scheduling the costs of Delta that the Beta system would be deployed as described previously and that exploratory research on the space-based components would continue through the 1990s; the costs of exploratory research are not included in our estimates, as such efforts are included already in the U.S. defense program. Full-scale development of the Delta system would begin in 2002 and be completed, for the most part, by 2010. The necessary launch capabilities and laser and battle management satellites would be acquired between 2008 and 2020. Deployment of the satellites might begin in 2016 and be completed in 2020, when the system would be fully operational.

(These dates obviously are arbitrary. An interesting alternative might be to deploy the Gamma system as described previously, while research continued on laser systems through the first fifteen to twenty years of the new century. At that point, a directed energy component might be deployed to replace the existing space-based component. Such a combination of systems would be considerably more expensive than any of our four notional systems, but it might be more capable.)

If deployed on the stated schedule, the Delta system would require roughly $11 billion per year to acquire and operate the

FIGURE 2: Annual Cost of Gamma System

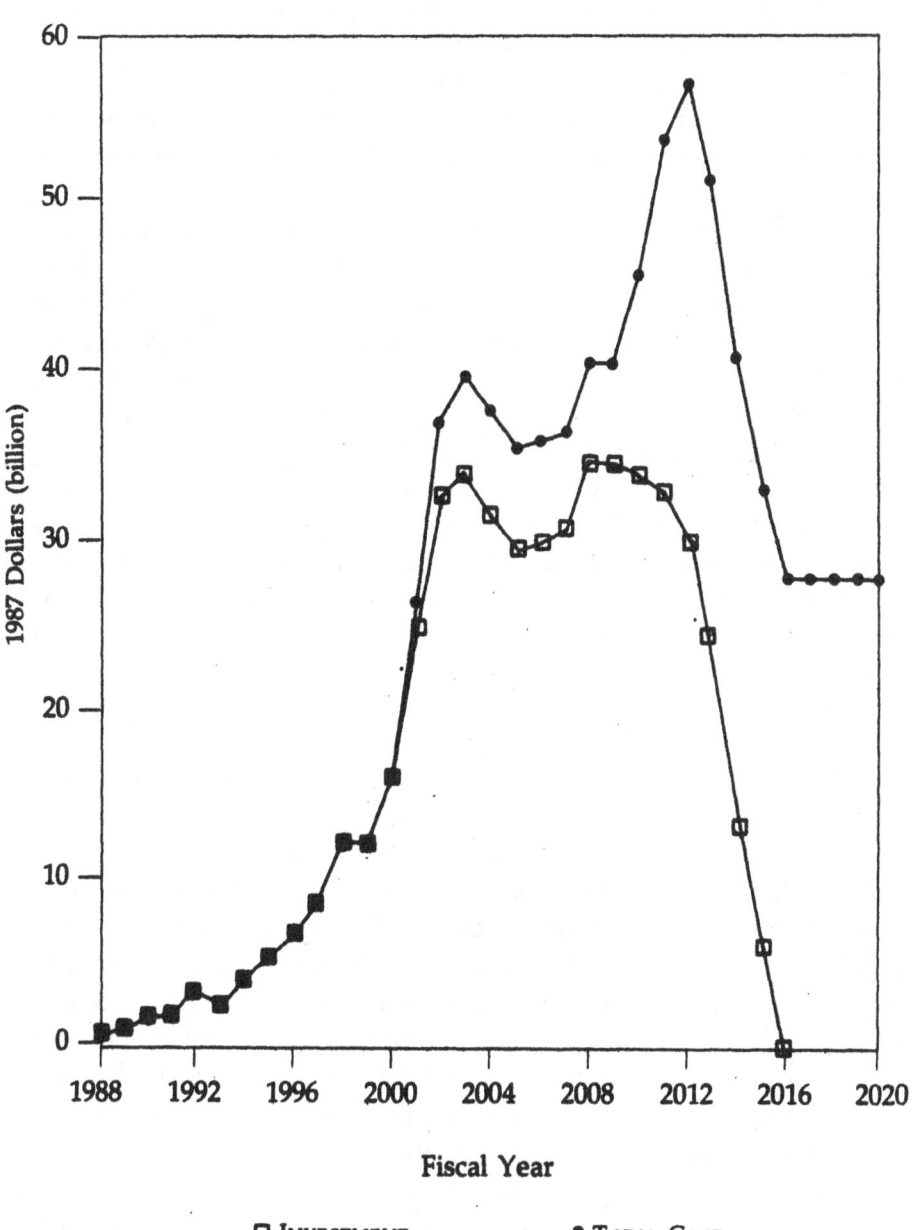

Fiscal Year

☐ Investment   ● Total Cost

NOTE: The figure includes the cost of thickening the system to defend against Soviet IRBMs.

shared components of the Beta system until roughly the year 2008, at which point appropriations would begin to rise. At its peak ten years of funding, 2011 through 2020, the Delta system would require annual appropriations on the order of $37 billion. Subsequently, it would require about $22 billion per year to operate the system. About two-thirds of the cost of the Delta system would allocated for investments.

The first thing that should be said about these figures is that the nation clearly could afford to deploy strategic defenses if it chose to do so. The most expensive system, Gamma, during its most demanding ten years of spending would entail annual expenditures averaging less than $50 billion. Such an allocation of the country's resources would represent a very significant national decision, but in the aggregate it would be well within the realm of the possible for a country as wealthy as the United States. In very rough terms, a decision to build a comprehensive strategic defense system like Gamma on the schedule we have described would entail a commitment to allocate between .5 and 1 percent of the nation's resources for the purpose of strategic defense each year and to sustain that commitment for at least fifteen years.

Having determined that the country could afford a comprehensive defense system does not take us very far toward a rational decision, however, as a second and more important question must still be answered: What would be the opportunity costs of a decision to deploy strategic defenses? What would the country have to give up in order to deploy strategic defenses? There is, indeed, no such thing as a free lunch; dollars spent for strategic defenses would not be available for other defense or federal civilian programs or for private uses. Scientists and engineers working on free-electron lasers could not work on other defense or civilian research programs, for example. Space launch capabilities used to deploy the Gamma or Delta systems would not be available for planetary exploration or military reconnaissance.

Of course, the pot of resources available to the nation is not fixed for all time. Economic growth will increase the nation's wealth and capabilities over time. Through the political process the citizenry can decide to increase or decrease the share of resources made available to the federal government or to alter the distribution of federal programs between defense and nondefense needs.

FIGURE 3: Annual Cost of Delta System

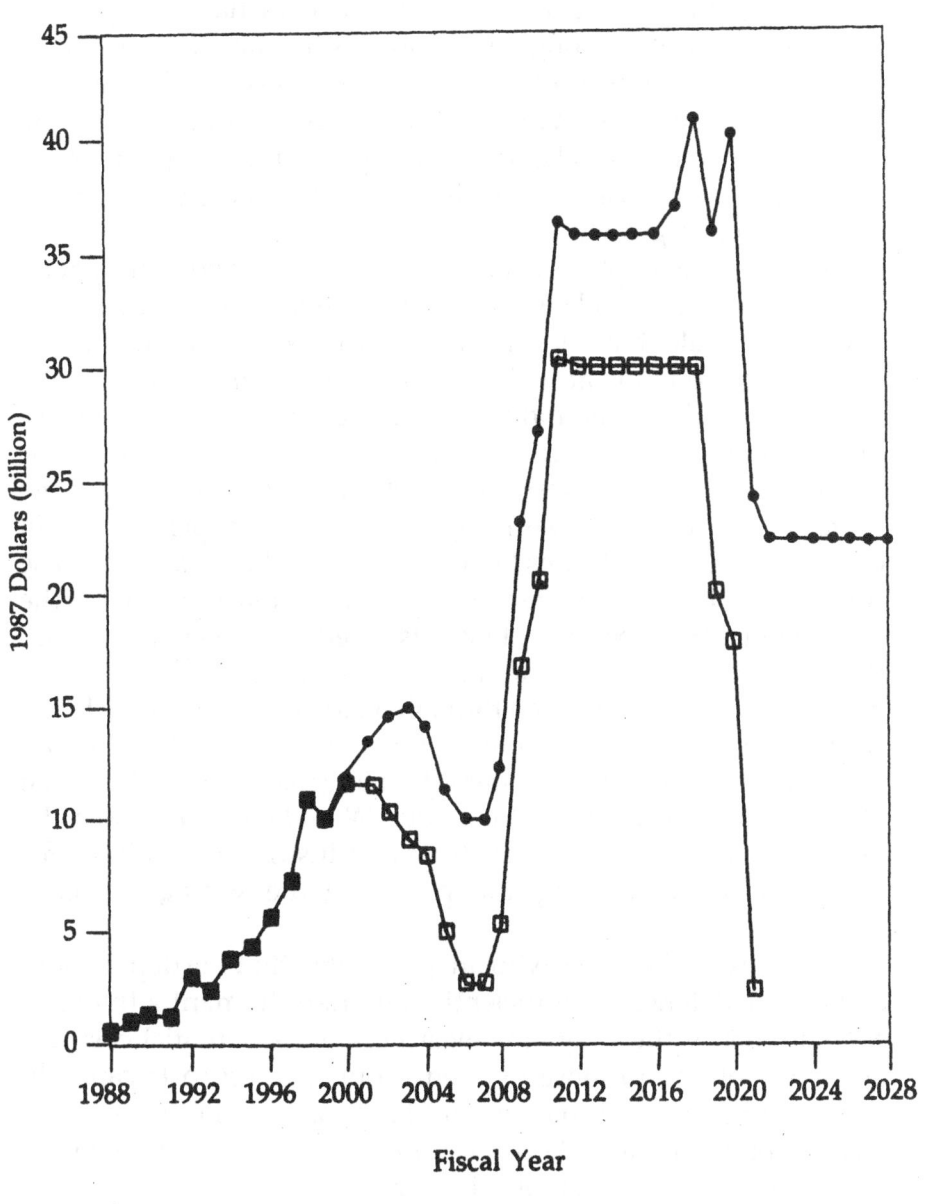

□ INVESTMENT   • TOTAL COST

NOTE: The figure includes the cost of thickening the system to defend against Soviet IRBMs.

The nation can also initiate programs intended to encourage or discourage private decisions to conform with national goals. It could, for example, through educational subsidies seek to expand the size of the scientific establishment and to shape the distribution of disciplines selected by individuals entering the scientific community. It could also take steps to improve the competitiveness of the nation's industries, thus increasing the prospects for more rapid economic growth.

A decision to deploy a comprehensive strategic defense system, because it would entail such a major commitment of the nation's resources, could have such effects itself—either stimulating or depressing economic growth, possibly restructuring the nation's spending priorities, and influencing the country's scientific and industrial sectors. All would depend, of course, on what economic and industrial policies were pursued during the period that the strategic defense system was being developed and deployed: Would other defense expenditures be reduced or would strategic defense costs simply be added incrementally to the defense budget? If the latter, would there be coincident cuts in federal civilian programs? If not, would federal revenues be increased to pay for the new program, or would it be financed through deficit spending? If the latter, would monetary policy be adjusted to avoid any inflationary consequences and to curtail any impact on interest rates and the value of the dollar in international markets? Would coincident steps be taken to compensate the scientific and industrial communities for the effects of developing and procuring the strategic defense system?

For those individuals who are persuaded that the deployment of strategic defenses is sufficiently important to merit a full-scale commitment to the program, regardless of its potential cost, examination of any proposed system's opportunity costs can help to determine how the country could best pay for the system. The analysis of potential trade-offs can inform decisions on the fiscal, monetary, and industrial policies that might reduce any adverse economic consequences of a decision to deploy strategic defenses and thus facilitate its implementation.

At the same time, for those who are undecided about the wisdom of pursuing the strategic defense initiative, consideration of opportunity costs should be an integral aspect of current decisionmaking.

From this perspective, it is essential in evaluating a proposed defense system to judge not only the degree to which it might work and its military and political consequences, but also what the nation would have to give up to acquire such capabilities. The key question is what trade-offs might be necessary between the deployment of strategic defenses and other defense or civilian needs.

In the remainder of this book we explore the potential opportunity costs of a decision to deploy strategic defenses, by examining such potential expenditures in four contexts: (1) the portion of the defense budget allocated to strategic forces, (2) the full defense budget, (3) the federal budget overall, and (4) the nation's economy as a whole.

7.
STRATEGIC DEFENSES IN THE CONTEXT OF HISTORIC SPENDING LEVELS FOR STRATEGIC FORCES

T HE MOST OBVIOUS WAY TO PAY for a strategic defense system would be to fund it within the strategic forces' portion of the defense budget. As will be seen, such an approach might be feasible for the Alpha or Beta systems but not for either of the comprehensive defense systems.

About three-fourths of the costs of strategic forces are funded within program I ("Strategic Forces") of the overall defense program. These costs, which include all research and development, procurement, and military construction appropriations associated with specific strategic weapons, as well as the direct operating costs of those military units that make up the strategic forces, are comparable to the costs included in the estimates in this book. Excluded from program I costs, and from our estimates, are the expenditures required to recruit, train, and support personnel assigned to strategic forces, much of the cost of reserve units that take part in strategic operations, most expenditures for intelligence and command and control systems and for basic and exploratory research. At present, for example, expenditures for the strategic defense initiative are not included in program I. The cost of developing and manufacturing nuclear warheads for strategic weapons, which are funded within the Department of Energy's budget, are also excluded from program I.[38]

In 1987 prices annual appropriations for strategic forces (program I) rose from about $16 billion in fiscal 1980, the last year before the initiation of President Reagan's defense buildup, to around $30 billion in fiscal 1984 and 1985; they have since declined to around $25 billion. Real annual growth on the order of 5 percent is projected by the administration for fiscal years 1988 and 1989.[39]

This level of expenditures on strategic forces is relatively high by recent standards but not so great when considered in the context of a longer period. Indeed, as Table 13 shows, spending for strategic forces has varied widely over the postwar years, depending primarily on the stage of U.S. efforts to modernize or expand its nuclear capabilities. All told, over this thirty-one-year period, annual appropriations for strategic forces have averaged about $29 billion in today's prices, roughly the current level of budgetary allocations. On average, the program has accounted for about 13 percent of the defense budget; the current level is only 8 percent.

Relatively speaking, strategic forces have been most important in budgetary terms in the 1950s, during the period of president Eisenhower's heavy reliance on nuclear forces through the doctrine of "Massive Retaliation," when the nation first developed and deployed nuclear-armed long-range missiles, modernized its force of strategic bombers, and constructed a substantial air defense system. During this period annual appropriations for strategic forces averaged $47 billion when expressed in 1987 prices, accounting for roughly one-fifth of all defense spending.

Appropriations for strategic forces dropped substantially during the Kennedy/Johnson years to about $35 billion annually, about 15 percent of total defense spending. This absolute and relative decline occurred despite the step-up in strategic modernization programs (Polaris, Minuteman) initiated by the Kennedy administration, largely because the program to acquire B–52 bombers was completed and the large number of medium-range B–47 bombers then in force began to be phased out.

Beginning in 1969 strategic appropriations dropped sharply, as the Nixon, Ford, and Carter administrations chose to live off the capital invested in strategic forces in previous years, thus husbanding resources for other defense and civilian needs. During these years the United States concentrated on improving the performance of existing bombers and missiles rather than building whole new systems. This trend bottomed out in fiscal 1979, when appropriations for strategic forces amounted to only $13.5 billion,

TABLE 13
Levels of Appropriations for Strategic Forces
(1946–1986)

Stage of Strategic Modernization (fiscal years)	Average Share of Total Defense Budget (percent)	Yearly Average ($ billions–1987)
1946–50	10.9	15.2
1951–60	21.5	47.4
1961–68	15.3	35.1
1969–80	8.6	18.1
1981–86	8.7	23.8
1946–86	13.3	29.0

Source: Derived from Kevin N. Lewis, *The Potential Large-Scale Budget Impacts of a Comprehensive Strategic Defense Effort: Some Parametric Analyses* (Santa Monica, Calif.: The Rand Corporation, P–7253, October 1985).

less than 7 percent of the defense budget. The trend toward greater budgetary allocations for strategic forces was initiated only in the final years of the Carter administration and then greatly accelerated by President Reagan.

Spending for strategic defenses, which is incorporated in these figures, declined sharply over this historic period. Appropriations for strategic defenses were a relatively large share of the strategic budget in the 1950s and early 1960s, when the United States deployed substantial defenses against Soviet bombers, but later in the period, and particularly in the early 1970s, they fell precipitously. Around this time existing plans to modernize the U.S. strategic air defense system with new interceptors, ground-to-air missiles, and radars were scrapped and existing forces cut back substantially. This period coincided, of course, with conclusion of the Anti-Ballistic Missile Treaty. Defending against the relatively small force of Soviet bombers made little sense in view of the nation's decision not to attempt to defend itself against the much larger and potentially more destructive force of Soviet missiles, but to depend, instead, on retaliatory capabilities to deter any attack.

Detailed information on spending for strategic defenses is not available but, according to one source, it accounted for only 14 percent of spending for strategic forces (about 1 percent of total defense spending) during the fiscal 1976–80 period. The Reagan administration has reinvigorated some air defense modernization programs but not substantially enough to raise these figures by any significant amount.

Given this history, would it be possible to fund the deployment of strategic defenses within the strategic forces budget as it traditionally has been circumscribed? The short answer is, "no"—unless three conditions were fulfilled: (1) if spending for strategic forces rose to the level common in the 1950s, between $40 and $50 billion in 1987 dollars, and (2) so long as there were no significant increases in current plans for modernizing strategic offensive forces, and, even then, (3) only if spending for strategic defenses did not exceed the levels we estimated to be necessary to deploy the Alpha or Beta systems. A comprehensive strategic defense system like Gamma or Delta could not be accommodated within the strategic forces portion of the budget even in the best of circumstances.

As we have noted, appropriations for strategic forces have declined somewhat from a peak around $30 billion in fiscal 1984 and 1985 to around $25 billion at present. This recent drop reflects the near completion of several programs that dominated the growth in strategic spending during the early 1980s, including most importantly the B–1 bomber and the MX missile. Further declines are not in prospect in the midterm, however, as budgetary requirements for the other strategic modernization programs are rising— the Advanced Technology Bomber, Trident submarines and missiles, and the Midgetman land-based, mobile missile, being the most important. Still other strategic programs, including cruise missiles programs, will continue to require major budgetary appropriations for many years. All told, requests for funding for strategic offensive forces can be expected at least to remain constant in real terms and more likely to rise at a real annual rate of around 5 percent well into the 1990s.

What might happen beyond that? In part, this could depend on whether or not the United States decides to deploy strategic defenses. In one scenario, requirements for modernizing strategic offensive forces might be reduced along with deployments of strategic defenses. The administration has stated its preference, if strategic defense technologies prove viable to negotiate a transition with the Soviet Union from defense postures that depend on the retaliatory capabilities associated with offensive forces to postures that depend primarily on defenses. Such a transition could involve major reductions in numbers of strategic offensive forces on both sides, as well as constraints on modernizing such forces, and thus result in reduced operating costs and much smaller appropriations for the development and procurement of strategic missiles and aircraft. In such a case there would be greater room within traditional levels of spending for strategic forces to develop and acquire strategic defenses.

Still, one would not expect such a regime to be negotiated successfully until the United States had demonstrated to the USSR through actions, its ability and willingness to deploy effective strategic defenses. To the extent that such an arms-control regime contributed to reduced budgetary pressures, therefore, it would be likely to occur toward the end of the deployment period included in our calculations, early in the next century. It might make feasible the development and acquisition of follow-on strategic defense systems within the traditional strategic forces budget, but not the initial systems considered here.

Moreover, an objective assessment based on experience would suggest that the negotiation of such a transitional arms-control regime would require such an extraordinary degree of cooperation between the United States and USSR as to be unrealistic. The record of U.S.-Soviet arms negotiations does not inspire confidence in the ability of the two countries to reach far-reaching accords. It is more likely that each side would both deploy strategic defenses and seek to overcome the opponent's defensive capabilities by improving its offensive forces.

This latter scenario could mean greater, rather than lesser, budgetary requirements for strategic offensive forces. Once the architecture of the opponent's defensive system was known, for example, each side might deploy new generations of missiles intended to exploit that system's weaknesses. We outlined several of these possibilities in describing the two comprehensive strategic defense systems, including missiles with reduced burn-times and missiles hardened against directed energy weapons. Each side might also seek to proliferate numbers of offensive forces and warheads, or to redeploy them in more concentrated areas, or simply to match increases in the opponent's forces for political reasons. All these factors could add measurably to spending for strategic offensive forces in the aftermath of a decision to deploy strategic defenses.

Thus, in contemplating how the United States might pay for strategic defenses, it appears to be imprudent to assume that a decision to deploy such weapons would be accompanied by reductions in budgetary demands for strategic offensive forces. At best, such reduced budgetary requirements might follow deployment of the initial defense system; at worst, deployments of defenses would lead to increased requirements for the modernization of strategic offensive forces.

In the absence of such reductions, funding the Alpha or Beta strategic defense systems within the strategic forces' portion of the defense budget would require increasing the level of program I in the 1990s to near its funding level in the 1950s—between $40 and $50 billion per year. Assuming that this level was sustained, it would be possible in very rough terms to allocate about $10 billion per year for strategic defenses and still continue to modernize strategic offensive forces as currently planned.

This amount would be just about enough to accommodate deployment of either the Alpha or Beta system. It would not be sufficient to fund a comprehensive defense system, however, nor would it be large enough for a limited system if, for whatever reason, there were further significant increases in requirements for modernizing strategic offensive forces. In the latter case, deploying even a limited strategic defense system while constraining spending for strategic forces to historic levels would require trade-offs between offensive and defensive requirements.

In its cost implications, constructing one of the limited strategic defense systems would be equivalent to adding a new component to the U.S. strategic posture. Total investment required for the Trident program, including 20 submarines and two generations of missiles at 1987 prices, will probably come to around $80 billion over a twenty-five to thirty-year period. The investment necessary to develop and construct the U.S. bomber force that will operate in the first decade of the next century, including 100 B–1 bombers, 144 Advanced Technology Bombers, and the cruise missiles and other armaments with which they will be equipped, will probably come close to $100 billion at 1987 prices, with expenditures spaced over a twenty-year period. Total investment for the Alpha system would be around $100 billion, spread over an eighteen-year period. If constraints on strategic spending necessitated trade-offs with offensive forces in order to deploy strategic defenses, this is the magnitude of what might have to be given up.

Spending $40 to $50 billion per year on strategic forces, moreover, would mean sustaining at least a 60 percent real increase in program I funds in the 1990s in addition to the 67 percent increase that has already taken place in the 1980s. If total defense spending remained constant in real terms over the deployment period, it would suggest an increase in the strategic forces' share of the overall defense budget from its low of 7 percent in 1979 to perhaps 14 to 16 percent in the mid-1990s. What might have to be given up to accommodate such growth in spending for strategic forces without raising the overall level of defense appropriations is examined in the next section.

8.
STRATEGIC DEFENSES IN
THE CONTEXT OF THE
OVERALL DEFENSE BUDGET

LIKE SPENDING FOR STRATEGIC FORCES, the defense budget has exhibited considerable volatility over the postwar period. Defense spending was kept very low immediately following World War II, rising sharply, but only briefly, to accommodate the Korean War. Once that conflict ended, the Eisenhower administration cut back sharply on defense appropriations in keeping with its overall philosophy that the greatest internal threat to the nation's security was fiscal irresponsibility and that nuclear weapons could substitute for more expensive conventional capabilities in defending against external dangers. The defense budget began to increase in the 1960s with president Kennedy's buildup, primarily for conventional forces, and rose steeply with the deepening U.S. involvement in the Vietnam War. It declined with the withdrawal from Southeast Asia, remained relatively low through most of the 1970s, and rose steeply again during the first half of the current decade.

The rises and falls in the defense budget over the full postwar period are pictured in Figure 4, in both current and constant prices. Even if incremental expenditures for the Korean and Vietnamese wars are factored out, the overall trend has been upward, with real growth in defense spending unrelated to the two conflicts averaging somewhat less than 2 percent each year.

Defense spending, however, has not risen nearly as fast as overall federal spending, nor has it kept pace with the nation's overall economic growth. As a share of total federal outlays, defense

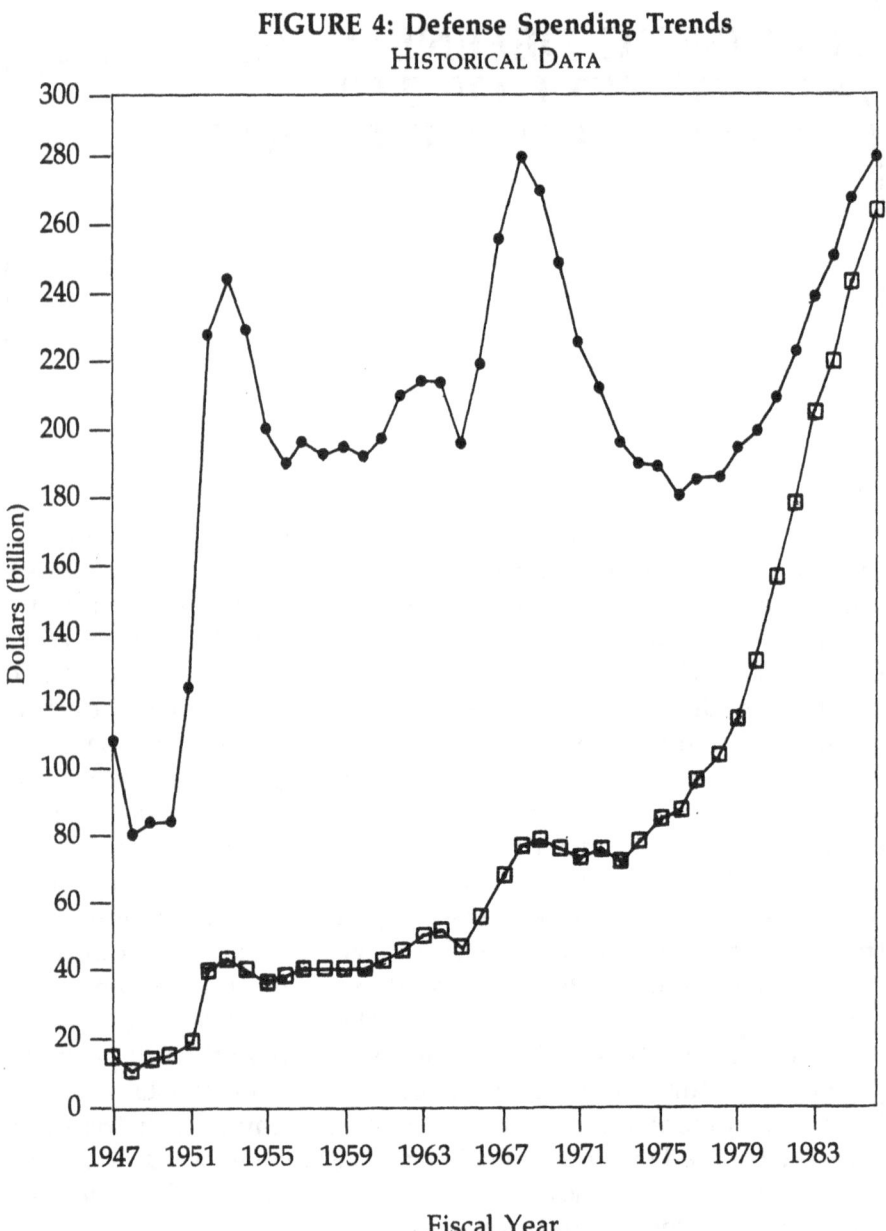

FIGURE 4: Defense Spending Trends
HISTORICAL DATA

Dollars (billion)

Fiscal Year

□ CURRENT $ • 1987 $

spending has declined from an average in excess of 50 percent between 1945 and 1965 to a figure around 25 percent during the past twenty years. This reflects, of course, the great expansion of domestic entitlement programs and growth in the federal government's interest payments, rather than any absolute cut in defense appropriations. The attempted restructuring of the nation's spending priorities during the Reagan administration has altered this trend, but not in a major way. Defense spending as a share of federal revenues increased from 22.5 percent in fiscal 1980 to 26.4 percent in fiscal 1986.

As a share of the nation's overall resources, defense declined fairly steadily from the 1950s, when—with the exception of the Korean War—it accounted for about 9 percent of the nation's resources, through the 1970s, reaching 4.8 percent of GNP in fiscal 1978 and 1979; the Vietnam period provided a temporary deviation from this trend, of course. The trend was reversed in the 1980s, and defense spending as a share of GNP rose swiftly to around 6 percent, where it now seems to have stabilized.

What is the likely future trend? The picture is clouded by the current dispute between the president and the Congress over budget priorities and by uncertainties about the effects of the Gramm-Rudman-Hollings deficit reduction legislation. In submitting the fiscal 1987 budget, the president proposed to increase the defense budget by about 5 percent in real terms in the coming year and projected 3 percent real annual growth through fiscal 1991. Judging from the budget resolution approved in June by the House/Senate conference committee, however, it appears that defense appropriations will not be permitted to grow in real terms in fiscal 1987; barring some major international crisis, this situation seems likely to prevail at least through fiscal 1991, by which time a balanced budget is supposed to be achieved.

Even with the prospect of constant budgets in the near term, however, the nation's historic proclivity to sustain modest real increases in defense spending and the virtual certainty of at least modest economic growth in the future, should result in rising defense budgets over the longer term. Three such possibilities are shown in Figure 5.

The low projection assumes that defense spending will increase on average over the next twenty-five years, in real terms, at about

FIGURE 5: Defense Spending Trends
ALTERNATIVE PROJECTIONS

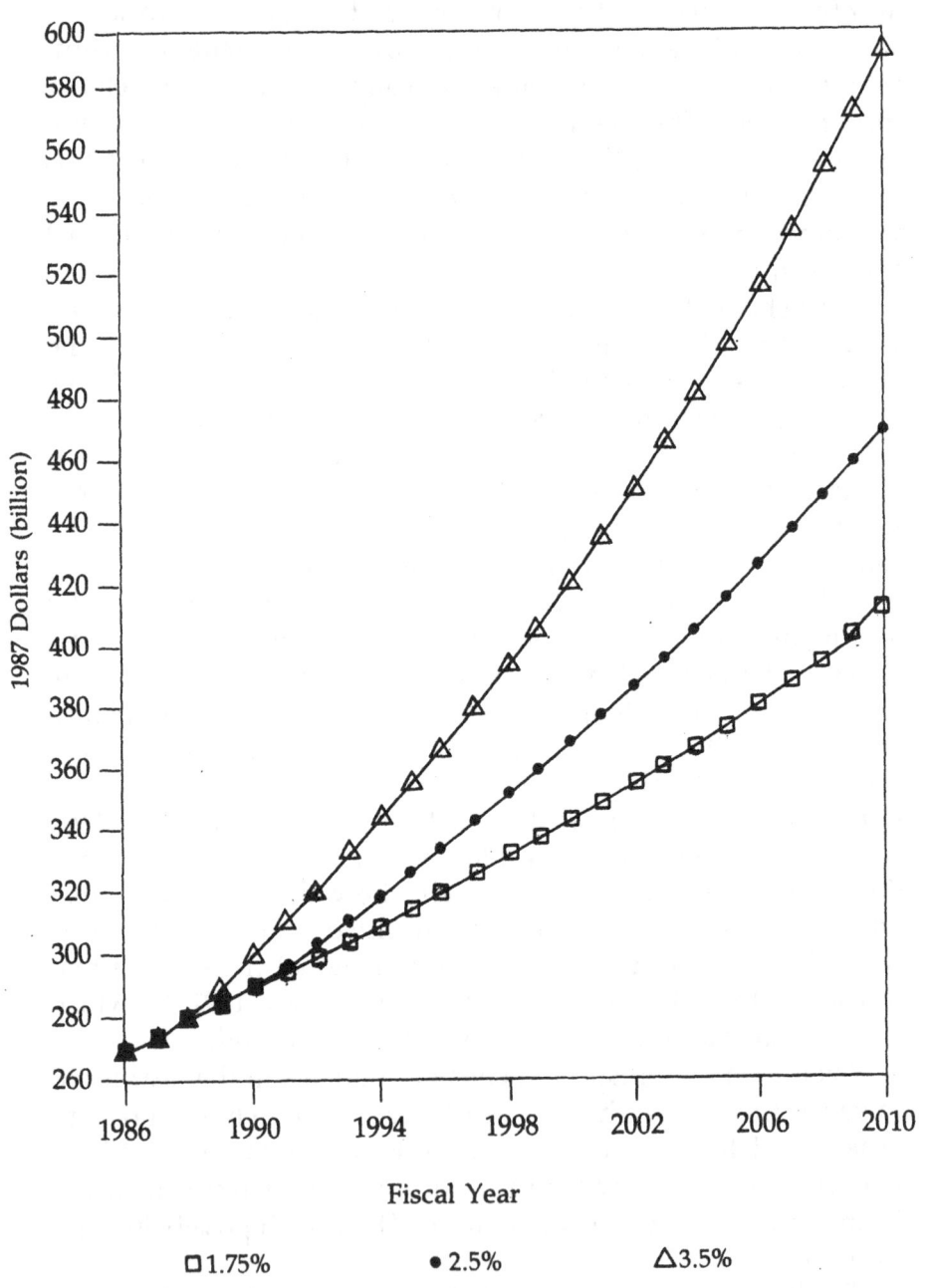

the same rate at which it has grown—excluding the effects of Korea and Vietnam—over the past forty years. (We used 1.75 percent annual growth in making the projection.) At such a pace defense outlays could be roughly $413 billion in the year 2010, $145 billion or 54 percent larger than they were in fiscal 1986. Even with such growth, defense would probably account for smaller shares of both federal expenditures and the nation's overall resources than it does today.

The midrange projection shows one logical outcome of the current congressional process. It assumes that for the period between fiscal 1986 and 1991 the nation's economy and federal revenues would increase as projected by the Congressional Budget Office in its February 1986 report, but that—as required by the Gramm-Rudman-Hollings legislation—federal spending would be gradually reduced in order to eliminate the federal deficit. We have assumed further, also in accord with Gramm-Rudman-Hollings, that the necessary reductions in federal spending would be taken proportionately in all programs. Beyond 1991 we have assumed that the nation's economy would grow annually at an average rate of 2.5 percent, that federal revenues would freeze at their projected 1991 level (19 percent of GNP), that federal spending would not exceed revenues, and that defense spending would remain proportionate to other federal programs, meaning that it would increase only with growth in GNP. In this scenario defense outlays would amount to $468 billion in 2010, an increase of $200 billion (75 percent) as compared to fiscal 1986; defense's share of total resources would be roughly 5.7 percent as compared to the current 6.2 percent.

The high projection shows the trend defense spending would follow if the president's program was fully implemented between now and fiscal 1991, and if defense's share of the nation's overall resources was thereafter held constant at 6 percent. It assumes further a 3.5 percent real growth rate in GNP. This means that domestic programs would have to take disproportionate cuts to reduce the deficit and that defense would rise to about 33 percent of the total federal outlays—its level in the 1960s. At the end of the period the defense budget would total $594 billion at 1987 prices, more than doubling fiscal 1986 spending.

The wedges under the three curves represent the incremental funds that could be made available for new defense initiatives.

Would the growth projected in any of the three be sufficient to absorb the incremental costs of developing and deploying any of the four systems described earlier in this book? On a superficial level the answer to this question is clearly, "yes," even in the case of the lowest growth projection. In principle, normal growth that can be anticipated in the defense budget should be sufficient to fund the development and procurement of even a comprehensive strategic defense system.

This first-order analysis neglects the possibility of competing demands on the defense budget, however. Strategic defenses will certainly not constitute the only claimant for anticipated growth in defense appropriations. Defense spending has tended to increase in real terms in the past because the nation has perceived rising needs for U.S. military power and because the real cost of acquiring and maintaining any given quantity of military capability has tended to rise throughout the postwar period. Any real growth in defense spending allocated for strategic defenses would not be available to fulfill these other requirements. A complete answer to the question of the adequacy of predictable growth in defense spending for the cost of strategic defenses, therefore, depends on a number of other assessments concerning the prospects for additional or reduced demands on the aggregate defense budget.

Most basically, as technology advances, the capabilities of individual weapon systems improve measurably; so too, however, do the salaries and training costs of the individuals who operate them, the cost of maintaining the systems, and the investment necessary to replace them. Secretary of Defense Caspar Weinberger maintains that it would require $30 billion each year simply to maintain the current stock of defense equipment; improving the capabilities of that equipment to take advantage of advanced technologies and to keep pace with improvements in the capabilities of potential adversaries would require much larger sums.[40] A second problem is associated with the declining pool of workers in the United States; recruiting enough people of sufficient quality to operate and maintain today's technically advanced military equipment is likely to require additional expenditures for salaries and other incentives.

Estimates of the annual real budget growth necessary simply to maintain the current force posture—without increases in force levels or additional missions—typically range between 2 and 4 percent. In the absence of such added expenditures, according to these analyses, either the size of the forces or their readiness for combat would have to decline. As we have noted, annual real growth in U.S. defense budgets has averaged less than 2 percent in the past and, indeed, U.S. combatant force levels have tended to decline over much of the postwar period.

One factor that might mitigate the demand for incremental defense expenditures simply to maintain current forces are the reforms now being instituted to streamline the weapons acquisition process and in other ways to reduce unit costs. But the potential here is limited; reformers would be delighted if these measures managed to reduce acquisition costs by 20 percent, meaning savings on the order of $25 billion at today's levels of expenditures.

Aside from these internal determinants of costs, world conditions will determine whether demands for U.S. military capabilities, and thus defense spending, will rise or fall. Conflicts and threats to Western interests in the Third World have been rising, for example, and show little evidence of subsiding in the future. They must be considered likely to constitute a continuing demand for rising defense budgets.

If U.S. allies increased their spending on defense, on the other hand, it would be possible for the United States to decrease its spending, but this seems unlikely, with the possible exception of Japan. If a breakthrough was made in relations with the Soviet Union and far-reaching arms-control agreements governing forces in Europe, this too could lead to reduced demands for U.S. military power, but again there is certainly no empirical basis for predicting such events.

The U.S. decision to deploy strategic defenses itself would affect this aggregate demand on U.S. military capabilities and overall defense budgets. On the negative side, a need to improve certain capabilities for the defense of NATO would follow logically from a decision to deploy comprehensive defenses; the possible cost of such initiatives was estimated in connection with, but not included in, the cost of the Gamma system; it amounted to around $160 billion.

On the positive side, if it was possible to reach the sort of arms-control regime governing a transition to a defense-oriented strategic posture, as previously described, it should also be possible to negotiate the type of settlement in Europe that would make possible substantial reductions in U.S. forces. As we have pointed out, however, both measures presuppose an extraordinary degree of U.S.-Soviet cooperation.

Deployment of a comprehensive strategic defense system might reduce overall demands for defense spending in one other way. By forcing the Soviet Union to allocate additional funds and, perhaps more important, scarce scientific and engineering resources for these purposes, strategic defense deployments conceivably could force reductions in Soviet spending on other types of forces and, thereby, decrease over time the rate at which the United States would have to improve its military capabilities. Any such effect seems likely to occur only over an extremely protracted period of time, however, and experience would suggest that the USSR would simply allocate whatever is necessary for strategic defenses incrementally; its command economy and authoritarian political system make the restructuring of national priorities far easier than such processes are for democracies.

It is, of course, impossible to predict what the net result of these contradictory possibilities might be. It does seem safe to forecast that demands for real growth in defense spending, even without a decision to deploy strategic defenses, are likely to exceed the 1.75 percent average annual growth rate assumed in our low projection. This suggests that if the defense budget grew at only its historic rate, expenditures for strategic defenses could be made only at the cost of not fulfilling other defense needs. This brings us back to the issue with which we ended the previous section: the potential trade-offs between deploying strategic defenses and other defense programs.

If defense spending was held to these historic levels, the opportunity costs of deploying strategic defenses in terms of other military forces could be great. For $160 billion (the ten-year systems cost of the Alpha system, for example) the nation could build 8 aircraft carrier battle groups and operate them for ten years. Alternatively, the ten-year systems cost of the Alpha system is the equivalent of the ten-year systems cost of 27 wings of F–15s or 14 armored divisions.[41]

To look at the most expensive case, deploying the Gamma comprehensive defense system would require a national commitment, in rough terms, to sustain annual expenditures between $35 billion and more than $50 billion for a period of fifteen years, and then nearly $30 billion per year indefinitely to operate the system. To put these figures in some perspective, consider that over the past five years the United States has appropriated in fiscal 1987 dollars, on average, $30 billion for all defense research and development accounts. Total procurement for all three military services has been running about $94 billion per year. If $40 billion per year was allocated for strategic defenses, roughly the amount necessary during the peak ten years of funding for the Delta system, this one mission would cost about the same amount as the navy's or air force's total investment budget in fiscal 1986 and about twice as much as the army invested that year.

To the degree that the country was willing to increase defense spending beyond the postwar experience, these potential opportunity costs in terms of other kinds of defense capabilities would be eased. The rate shown for our highest projection of total defense spending, for example, would seem great enough to accommodate both the costs of strategic defenses and other likely demands for increased defense spending. Increasing defense spending for a sustained period at an annual rate of 3.5 percent, however (as is assumed in the high projection), given likely economic trends, would require national decisions both to reverse the long-standing trend toward a reduced burden of defense spending on the nation's overall resources and to restructure priorities in federal spending. These potential opportunity costs are examined in the next sections.

9.
STRATEGIC DEFENSES IN THE CONTEXT OF TOTAL FEDERAL EXPENDITURES

IN CONSIDERING THE OPPORTUNITY COSTS of strategic defenses in the broader context of total federal spending, the essential choices are the same: either other federal programs would have to be given up in favor of deploying a strategic defense system or federal spending would have to be increased beyond anticipated levels to accommodate the new program. Let us consider each possibility in turn.

Most federal civilian programs do not begin to approach the prospective cost of deploying strategic defenses. The Tennessee Tom-Big-Bee Waterway, for example, one of the largest recent projects of its kind, cost less than $3 billion at 1987 prices over a fourteen-year period. Similarly, the Santa Ana River mainstream project, one of the largest current irrigation projects, is expected to cost less than $2 billion at 1987 prices.[42]

Somewhat more comparable have been the nation's largest space efforts. Program Apollo, which successfully put U.S. astronauts on the moon, cost about $66 billion over a twelve-year period when expressed in fiscal 1987 prices. This would exceed, but not be dissimilar to the cost of the missile defense component of the Alpha system, excluding the component that would be used to defend against bombers and cruise missiles. Closer to the full cost of the Alpha system would be the expense of rebuilding the nation's interstate highway system. From its start in 1956 through 1985, a thirty-year period, the current federal road network cost about $140 billion when expressed in fiscal 1987 prices.

Close to the full cost of the Gamma system would be implementation of the recommendations of the president's Commission on Space. This blue-ribbon panel proposed in early 1986 that the nation embark on a major and sustained effort to build a staffed outpost on Mars. As a first step, the plan would establish a permanent settlement (and manufacturing facility) on the moon early in the new century. The panel estimated the cost of this ambitious program to be $700 billion through the year 2020.[43]

More generally, the civilian programs that begin to approximate the potential cost of comprehensive strategic defenses are those that involve substantial expenditures on a continuing basis. Agricultural subsidies from 1945 through fiscal 1986, a forty-two-year period, for example, will have accounted for about $330 billion in federal outlays when expressed in fiscal 1987 prices. Medicare provides a closer analogy: from its inception in 1967 through 1986, a twenty-year period, the nation has paid, in fiscal 1987 prices, only 10 percent more for Medicare than we estimated would be required for the Gamma system—$770 billion.[44]

The relative magnitude of potential annual expenditures for strategic defenses as compared to federal civilian programs can perhaps be seen most clearly in Table 14, which lists both the average annual peak ten-year funding levels for the notional strategic defense systems and the actual expenditures for a sampling of federal programs during the current fiscal year. It should be clear from these relative magnitudes that increasing the defense budget to accommodate the cost of strategic defenses, without also raising the level of overall federal spending, would necessitate extremely difficult choices between defense and civilian programs. Indeed, the experience of the first five years of the Reagan administration suggests that this strategy for financing strategic defenses simply would be impracticable.

During these years, a concerted effort on the part of the administration and its allies in the Congress to restructure national priorities reduced the portion of federal spending accounted for by nondefense programs from 77.5 percent in fiscal 1980 (the last complete Carter budget) to 73.6 percent in fiscal 1986. This shift would amount to $35 billion at the level of spending projected for fiscal 1987, more or less enough to fund three of the four defense systems. Indeed, these figures mask the degree to which domestic

TABLE 14
Average Annual Expenditures During Peak Funding for
Notional Strategic Defense Systems and Federal Outlays
in 1986 for Selected Civilian Programs

Program	Amount ($ billions–1987)
Water Resources	4
Community Development	5
Higher Education	9
Alpha System	10
Beta System	11
Ground Transportation	19
Farm Income Stabilization	25
Medicaid	26
Health Care Services	30
Delta System	37
Gamma System	44
Medicare	71
Social Security	208

Source: OMB, *Historical Tables, Budget of the United States Government*, FY 1987.

programs have been reduced, as interest payments increased substantially during this period and are included in the nondefense percentages just cited. If net interest charges are omitted, it appears that nondefense federal spending declined by around $80 billion in real terms during this period—more than enough to fund any of the notional systems.

Repeating a shift in federal spending priorities of comparable magnitude to accommodate a new increase in defense budgets would appear to be an unlikely prospect, however. The potential base for such reductions is no longer large. The largest portion of nondefense spending (60 percent) is used for entitlements, like Social Security and other mandatory spending programs. As David Stockman apparently learned to his dismay, there is virtually no support for cuts in any of these programs. Net interest payments currently account for another 18 percent of nondefense spending. Factoring out these expenditures leaves around $180 billion (in fiscal 1987 prices) for annual, discretionary, nondefense spending. Funding the Delta system in its peak years solely by reducing these expenditures would require cutting one-fifth of discretionary nondefense programs; funding the Gamma system in this way would mean cutting one-fourth of these programs.

The vast majority of voters appear to believe that those nondefense programs that could and should be cut have already been terminated. At least most Representatives and Senators perceive that to be the public's belief. For the past two years the Congress has turned down virtually all the administration's proposals to end domestic programs. In the fiscal 1987 budget process, defense is being forced to take at least proportionate cuts to domestic programs; if anything, nondefense spending may increase relative to defense spending over the next few years.

Federal revenues will increase, of course, along with economic growth, permitting increases in federal spending without the accumulation of new deficits. If the nation's economy grew at an average long-term annual rate of 2 percent, for example (a figure on the conservative side of most economic projections) and federal revenues remained at their current 19 percent of national resources, annual federal outlays could be 50 percent larger in the year 2006 without incurring additional deficits. Assuming that defense retained only its current share of federal spending, thus obviating any need for cuts in nondefense programs, defense spending could still be $110 billion greater in 2006 than it was in fiscal 1986—an amount that could be allocated for either the types of demands for incremental defense spending described previously or for strategic defenses or that could be apportioned between them.

There will be many other claimants on future increases in federal revenues, however. Some experts have argued, for example, that for many years the nation has not invested sufficient amounts to maintain its economic infrastructure adequately. Pressures are likely to rise in the future for added federal subsidies for roads, airports, sewage treatment plants, and the like. Others have argued that the federal government should be investing more heavily in the nation's educational system and in civilian research and development in order to maintain the U.S. competitive position in modern industrial technologies and basic sciences.

The basic national demographic trend—the gradual increase in the average age of the U.S. population—will both pose new demands for nondefense spending and constrain the potential growth in federal revenues. An aging population means that the fiscal requirements of many entitlement programs—Social Security and Medicare being the most demanding—will increase more rapidly. An aging population, moreover, means that the ratio of the active work force to the community of retired individuals is likely to diminish. Such a trend could reduce the prospective size of federal revenues and, therefore, assuming that a balanced budget is achieved and maintained, diminish the increase in federal spending that might otherwise be made possible by economic growth.

Although we are not in a position to carry out a quantitative projection of likely long-term trends in federal revenues and nondefense spending, it does seem clear that there will be many claimants for whatever increase in federal spending becomes possible as a consequence of economic growth. Together with likely constraints on the growth in federal revenues and barring a protracted international crisis or war, these factors seem likely to limit annual real growth in the defense budget, on average, to no more than its historic rate of less than 2 percent. At this rate, growth in defense spending would not be sufficient to fund the development and deployment of a comprehensive strategic defense system without compensating reductions in other military forces.

The United States, therefore, faces some difficult decisions. If it wishes to deploy a strategic defense system, yet retain the federal government's share of the economy at its projected level, it will have to give up something in return. Candidates would include the types of cuts in nondefense discretionary programs that Congress has rejected over the past five years, reductions in entitlement programs—a possibility rejected by both the president and the Congress, and sizable cuts in other types of military forces.

One obvious alternative would be to increase federal revenues beyond the levels that can be anticipated as a result of economic growth; in other words, to increase the government's share of the nation's resources in order to accommodate the fiscal requirements of strategic defenses along with the other defense and nondefense demands on the federal budget. A second alternative would be to finance strategic defenses by permitting federal deficits to increase at the same time as the defense system was being constructed—a device that would, really, postpone the problem rather than face it. Both possibilities are examined in chapter 10.

10.
STRATEGIC DEFENSES IN
THE CONTEXT OF THE
NATION'S ECONOMY

$\mathbf{A}$s WE HAVE NOTED, in very rough terms, deploying a comprehensive strategic defense system would require the commitment of roughly .5 to 1 percent of the nation's resources for a period of about fifteen years. This would be feasible for the country; even larger efforts have been undertaken in wartime. Initiating such an allocation of national resources could impact adversely on the economy, but the economic consequences of such major expenditures would depend on a wide range of controllable and uncontrollable factors. Consider two examples.

From fiscal 1965 to 1969 defense spending increased from $51 to $83 billion, a nearly one-third increase in real terms, as the nation escalated its involvement in Vietnam; this rise increased defense's share of GNP by 1 full percentage point (6.8 to 7.8 percent). This increase in defense spending, moreover, took place at the same time as other parts of the federal budget were growing rapidly; total federal outlays increased from 17.6 to 19.8 percent of GNP over these five years. Additionally, president Johnson chose to permit these spending increases without raising taxes. Together, these actions had major economic effects: the overall expansion in government spending, defense and nondefense, stimulated a surge in output that cut the unemployment rate substantially but also resulted in major inflationary pressures.

Between 1979 and 1983, on the other hand, defense spending rose from $116 to $210 billion—also a real increase of about one-third over a five-year period—as the nation acted to correct what it perceived to be a decline in U.S. military capabilities relative to those of the USSR. Although federal nondefense spending was constrained during this period, tax revenues were cut sharply. Despite the fact that defense spending as a percentage of GNP increased by 1.5 percentage points (4.7 to 6.2 percent), a larger relative increase than the Vietnam case, opposite economic trends obtained: GNP growth slowed, unemployment remained persistently high, while inflation subsided impressively.

Thus, in two situations analogous to the magnitude of increased defense spending that might be associated with the deployment of comprehensive strategic defenses (although they applied to only one-third as sustained a period of time), the near-term macroeconomic consequences of defense budgetary decisions were dominated by other dimensions of federal policy. Although overall federal spending was more stimulative in the early 1980s than in the late 1960s (total federal outlays as a share of GNP rose 3.8 points during the early 1980s, as compared to 2.2 points in the late 1960s), primarily because of the coincident cut in federal taxes, monetary policy was more restrained in the 1980s than in the 1960s. This seems to have overwhelmed any potential effects of the defense run-up on inflation. The strength of the dollar relative to other currencies in the early 1980s also was important in determining the economic situation, as it depressed U.S. exports and encouraged imports, thus contributing to the poor employment record. In short, the immediate effects of changes in defense spending—even of the magnitude envisioned to deploy strategic defenses—are likely to be overwhelmed in their macroeconomic consequences by other relevant considerations.

If we assume for the moment that because of existing defense and nondefense demands on any prospective increases in federal revenues, strategic defenses should be considered an incremental requirement for federal expenditures, a key question is whether federal revenues would be increased further to pay for the system. Financing comprehensive strategic defenses, like the Gamma system, on such a "pay-as-you-go" basis might avoid many near-term adverse macroeconomic effects, such as increased deficits or greater inflation.

Such a "pay-as-you-go" policy would require, however, an increase of about 5.6 percent in the current annual level of federal revenues. This amount might be raised by increasing revenues from individual income taxes by about 11 percent. According to The Brookings Institution's "Tax Model," in 1986, such a raise would mean, on average, an increase of about $570 in the yearly federal tax bill of a family earning between $30,000 and $50,000 per year. A family earning between $20,000 and $30,000 would pay, on average, an additional $260 in federal income taxes each year.[45] Alternatively, under current tax codes, revenues from corporate income taxes could be increased by 50 percent, or the income from excise taxes more than doubled, or some combination of new taxes could be installed.

Because federal revenues will increase as a result of economic growth regardless of such possible surcharges to pay for strategic defenses, the relative incremental tax burden would be smaller at the time a strategic defense system actually would be acquired. If we assumed a 2 percent average annual real increase in federal revenues, for example—the net effect of 2.5 percent annual economic growth and a decline in the working population— financing the Gamma system in the first decade of the next century would require an increase of 3 to 4 percent in federal revenues each year, as compared to the 5.6 percent for the federal tax base in 1986. Given the Congress's tentative revision of the tax code, we decline to estimate the consequences of such a surcharge on individual taxpayers.

Of course, any increase in federal taxes would mean reduced individual consumption and greater government spending. The effects of such changes on the prospective rate of economic growth, on employment, and on inflation would depend on the specifics of the tax plan, the specifics of the expenditures for strategic defenses, and many other factors. If tax revenues were not increased to pay for the strategic defense system, the fiscal and monetary policies established to help contain any adverse economic effects would be of first-order importance. Other factors, some of which are relatively uncontrollable in the short term, such as trends in the cost of basic commodities, the international competitive position of U.S. industries, and the relative strength of national currencies, also would play determining roles.

In short, if the nation decided to deploy a comprehensive strategic defense system and to increase the magnitude of the federal budget to pay for it, it would appear that any potentially adverse economic consequences could probably be minimized by establishing additional taxes and by setting fiscal and monetary policies to take account of these new circumstances. (This assumes that uncontrollable economic circumstances were not severely negative.)

Greater difficulty might be encountered in seeking to contain the effects of such a large federal program on specific industrial sectors. The consequences we have described of the major research projects and huge procurement programs for the electronic and aerospace industries, for example, could be quite severe if compensating measures were not taken in a timely manner. Shortages of trained engineers, computer programmers, and other occupations could result, or shortages could develop in supplies of certain types of materials and electronic components. Depending on the overall economic situation, these problems could lead to sharp increases in the price of certain types of civilian goods and to a slower rate of development of civilian technologies, at least in the near term, which could impact adversely on the nation's economic growth. These effects could be ameliorated over time, and perhaps ultimately reversed, if the technologies developed for strategic defense spun off advances in civilian technologies and industrial techniques. Assessing these prospective effects requires far more detailed studies of individual industries than we are prepared to carry out, however.

CONCLUSION

I N ALL LIKELIHOOD, if the nation does decide to build a comprehensive strategic defense system, it would be financed through a mixture of the various means we have discussed. Spending for other defense needs would be somewhat lower than it might have been in other circumstances, as would federal spending for nondefense programs. Individual, corporate, and federal excise taxes might be somewhat higher than they could have been in other circumstances. The rate of inflation might be slightly greater and the rate of economic growth perhaps a little lower.

Because the costs of the system would be distributed among these various means, they would not likely be so visible. But whatever the precise means adopted to pay for a strategic defense system, the dollars spent on it would be real. They would represent forgone opportunities for the nation, either as taxpayers, individually, or as a society, collectively. As a shorthand, the income tax surcharge mentioned above, $570 per year for the average family earning $30,000 to $50,000 annually (at the current level of federal revenues and under current tax laws) is probably the most direct way of understanding the opportunity cost of deploying a comprehensive strategic defense system. Whether such a tax would actually be imposed or strategic defenses financed through some other means, that figure represents the opportunity cost of the system.

In aggregate economic terms, it would be feasible for the United States to develop and build comprehensive strategic defenses—

assuming that the technical problems identified in the previous sections were solved and that costs were held to levels not significantly greater than those estimated in this book. Such an accomplishment could provide greater security to the nation and have other beneficial political and military effects. Despite the near- and mid-term cost, the project might, over the very long term, have beneficial economic effects as well, by stimulating the development of technologies and industrial techniques, which could improve the nation's economic productivity and competitive position in the world economy.

Still, the costs of the project would be substantial—regardless of their distribution. The issue for all citizens to ponder is whether the potential benefits of strategic defenses would be great enough to warrant these forgone opportunities for other private or collective enterprises.

NOTES

1. Testimony of General Daniel Graham (USAF-ret.) before the Defense sub-committee, Committee on Appropriations, U.S. House of Representatives, *Defense Appropriations for Fiscal Year 1985* (Washington, D.C.: GPO, 1984), 860.

2. Zbigniew Brzezinski, Robert Jastrow, and Max Kampelman, "Defense in Space Is Not 'Star Wars,' " *The New York Times Magazine* (January 27, 1985), 29.

3. Hans A. Bethe et al., "Letters to the Editor: The Bid to Shoot Down 'Star Wars,' " *The Wall Street Journal* (January 17, 1985), 27.

4. William D. Hartung et al., *The Strategic Defense Initiative: Costs, Contractors, and Consequences* (New York: Council on Economic Priorities, 1985), 45.

5. *The New York Times* (April 25, 1986), A36.

6. Ibid.

7. These data were adapted from information in Kevin N. Lewis, *The Potential Large-Scale Budget Impacts of a Comprehensive Strategic Defense Effort: Some Parametric Analyses* (Santa Monica, Calif.: The Rand Corporation, P-7253, October 1985).

8. What might constitute an attractive target for the Soviet Union clearly would depend on circumstances and personal judgments. Even if the two nations began with equal inventories of weapons, a Soviet leader might be willing to utilize several weapons to destroy a single U.S. weapon if he valued the expected reduction in damage to the USSR more than he valued the increased damage that the expended Soviet weapons might otherwise cause to the United States. At the same time, and so long as we assume balanced arsenals to begin, it does seem clear that neither side would be likely to continue disadvantageous exchanges of weapons for very long. Such exchanges eventually would result in the attacker having significantly smaller capabilities than its opponent to destroy primary targets: locations of political leaders, other types of military forces, industrial facilities, and so on. Consequently, an attacker pursuing disadvantageous exchanges would have eventually a significantly reduced means of deterring enemy strikes against its own primary targets. Our characterization of "disadvantageous exchanges" as militarily unattractive to the Soviet Union thus seems reasonable. An Alpha defense system of course could be designed to make strikes against the bases of U.S. strategic forces even less attractive, but it would be more expensive.

9. Airborne air defenses are not always cheaper than ground-based air defenses; relative costs depend strongly on the shape and size of the area being defended, the precise characteristics of the two proposed classes of air defenses, and the tactics chosen by attacking aircraft.

10. Day-to-day alert rates for U.S. strategic forces tend to be lower than these figures; see, for example, Bruce G. Blair, *Strategic Command and Control: Redefining the Nuclear Threat* (Washington, D.C.: The Brookings Institution, 1985), 311. In making our calculations we assumed that crisis conditions would lead the United States to maintain significantly higher rates. If the normal lower rates were assumed, it would mean that more weapons would not be ready to escape from their bases, meaning that making the bases militarily unattractive to attack would require stronger, and thus more costly, strategic defenses.

11. The numbers of Soviet launchers and their characteristics, including the numbers and types of weapons carried by each launcher, are taken from: U.S. Department of Defense, *Soviet Military Power*, 4th ed. (Washington, D.C.: GPO, 1985); International Institute for Strategic Studies, *The Military Balance: 1985–86* (London: IISS, 1985); and John Moore, ed., *Jane's Fighting Ships: 1980–81* (New York: Jane's Publishing Inc., 1980).

12. The less capable Patriot air defense missile is a rough analogue for the HEDI missile. According to the *Military Cost Handbook* (Data Search Associates, 1983), the 277 Patriot missiles purchased in 1983 cost $2.8 million each, a figure that would be approximately $3.3 million in 1987 dollars. If these 277 missiles are assumed, conservatively, to have been the first of the Patriot missiles produced and a 90 percent learning curve also is assumed, the first unit cost of the Patriot would be roughly $7 million in 1987 dollars. A second source has projected a first unit cost of roughly $25 million in 1987 dollars for the longer-range Spartan antiballistic missile interceptor. These two estimates thus suggest that our assumption of a $6 million first unit cost for the HEDI missile is conservative.

13. The 50 percent increase we have projected for the first unit cost of the LEDI missile, as compared to the HEDI, may be a little high. If we had assumed that the first LEDI would cost the same amount as we assumed for the first HEDI, $6 million, the total cost of the Alpha system would be less than 3 percent below the $160 billion total now estimated.

14. Several sources indicate that a ground-based terminal intercept radar might cost in the neighborhood of $100 million. The radar they describe apparently would operate without the assistance of relatively nearby airborne laser radar systems that we included in the Alpha system, however. If we had used a figure of $100 million for each radar, the total cost of the Alpha system would be $6.4 billion greater, an increase of about 4 percent.

15. The *Military Cost Handbook* gives a procurement cost of $35 million in 1983 dollars for the TR1–A. Accounting for inflation and allowing $10 million per aircraft for the required modifications yields a cost estimate of $50 million per aircraft in 1987 dollars.

16. This is the *Military Cost Handbook* figure, converted to 1987 dollars.

17. These estimates were provided by the Engineering and Services Division, Office of the Deputy Chief of Staff for Logistics and Engineering, U.S. Air Force.

18. The ''third generation'' shelters currently being built in Europe are hardened against only conventional attack; they would need additional features to protect against the more exotic effects of nearby nuclear explosions.

19. While it might seem logical to use twenty or thirty years of operating costs in estimating the total cost of a system that is expected to last for several decades,

such a procedure would ignore the fact that the current significance of future costs declines the further into the future such costs would be incurred. Including ten years of the full operating costs is generally considered a reasonable way to capture the current significance of a longer stream of discounted future costs.

20. These figures are based on data in Samuel Glasstone and Philip J. Dolan, eds., *The Effects of Nuclear Weapons*, rev. ed. (U.S. Department of Defense and U.S. Department of Energy, 1977).

21. This assumption would lead to overestimates of costs only if the two-layer boost-phase defense that we have adopted proved less cost-effective than the midcourse or midcourse plus boost-phase systems that might be employed instead. Our statements regarding the practicality of one approach versus the other are meant less as judgments as to what ultimately might prove feasible technically than as explanations for the particular choices we made for the sole purpose of constructing cost estimates for notional comprehensive defense systems.

22. This force is similar in size and structure to that described in U.S. Department of Defense, *Soviet Military Power*, 4th ed. (Washington, D.C.: GPO, 1985) and *The Military Balance: 1985–86*.

23. U.S. Congress, Office of Technology Assessment, *Ballistic Missile Defense Technologies* (OTA–ISC–254; Washington, D.C.: GPO, September 1985), 173.

24. The deployment area we have assumed is 3 million sq.km. in size, about one-third of the 10 million sq.km. assumed to be the size of the Soviet Union's current ICBM deployment area in Richard L. Garwin, "How Many Orbiting Lasers for Boost-Phase Intercept?" *Nature*, vol. 315 (May 23, 1985), 286.

25. We have assumed that the booster burn-time would be 90 seconds and that 10 seconds of the total 90 would be lost to the defense while it confirmed the Soviet attack and made the decision to launch defending interceptors. Eighty usable seconds of booster burn-time plus 20 seconds of postboost time during which an intercept would still kill nearly all of the RVs carried by the postboost vehicle yields a total of 100 seconds for the defending interceptors to cover the distance from their host satellites to their targets.

26. In U.S. Department of Defense, *Soviet Military Power*, 5th ed. (Washington, D.C.: GPO, 1986) the Soviet Union is credited with "ground-based lasers that could be used in an anti-satellite role today and possibly a BMD role in the future."

27. R.G. Finke, J.H. Ashmore, C.J. Donlon, R.C. Oliver, and R.S. Swanson, *Continuing Issues (FY 1985) Concerning Military Use of the Space Transportation System* (IDA Paper P–1889; Alexandria, Va.: Institute for Defense Analyses, December 1985), pages II-29 through II-40.

28. John Pike gives a unit cost of $24 million for the air-launched miniature vehicle (ASAT) in *The Journal of the Federation of American Scientists, Public Issues Report* (November 1983), 4. Informal discussions with analysts who have studied this program suggest that his estimate is high by a factor of about two.

29. *Ballistic Missile Defense Technologies*, 185 and Garwin, "How Many Orbiting Lasers."

30. Garwin, "How Many Orbiting Lasers."

31. Hans A. Bethe and Richard L. Garwin, "Appendix A: New BMD Technologies," *Weapons in Space*, vol. II ("Implications for Security"), *Daedalus* (Summer 1985), 343.

32. "Panel Urges Defense Technology Advances," *Aviation Week & Space Technology* (October 17, 1983), 16. The article describes a 25-megawatt mid-infrared chemical laser with a 100-second total run time and 15 m. rather than 10 m. optics. Our smaller optics should lead to a somewhat smaller total weight.

33. General Andrew J. Goodpaster et al., *Strengthening Conventional Deterrence in Europe: A Detailed Program for the 1980s*, Report of the special panel of the European Security Study (Boulder, Colo.: Westview Press, 1985).

34. "NATO Ministers Can't Abdicate CW Decision, Says SACEUR," *Jane's Defence Weekly* (April 27, 1985), 719.

35. William W. Kaufmann, "Non-Nuclear Deterrence," *Alliance Security: NATO and the No-First-Use Question*, edited by John D. Steinbruner and Leon V. Sigal (Washington, D.C.: The Brookings Institution, 1983), 43–90.

36. Ibid.

37. *Time* (June 23, 1986), 20–21.

38. A historical analysis of spending for strategic forces can be seen in William E. Dupuy, Jr. et al., *U.S. Historical Budgets and Strategic Force Levels: 1945 Through 1972 (U)* (J. Watson Noah Associates, Inc., FR-114-XOD, June 1975).

39. Data for program I and total defense appropriations in this section are taken from U.S. Department of Defense, *Annual Report of the Secretary of Defense*, various years (Washington, D.C.: GPO).

40. Caspar W. Weinberger, *Annual Report to the Congress, Fiscal Year 1987* (Washington, D.C.: Department of Defense, 1986).

41. Ten-year systems cost data were provided by the Congressional Budget Office.

42. Project costs were provided by the Committee on Public Works and Transportation, U.S. House of Representatives.

43. National Commission on Space, *Pioneering the Space Frontier* (Toronto, Canada: Bantam Books, 1986).

44. Budgetary data were derived from U.S. Office of Management and Budget, *Historical Tables, Budget of the United States Government*, various fiscal years (Washington, D.C.: GPO); supplementary data were supplied by various congressional staffs.

45. Data were derived from information provided by Harvey Galper of The Brookings Institution.